For the Love of Enzymes

For the Love of Enzymes

THE ODYSSEY
OF A BIOCHEMIST

Arthur Kornberg

HARVARD UNIVERSITY PRESS
Cambridge, Massachusetts
London, England
1989

Library of Congress Cataloging-in-Publication Data

Kornberg, Arthur, 1918–
    For the love of enzymes.

    Bibliography: p.
    Includes index.
    1. Kornberg, Arthur, 1918-   2. Biochemists—
United States—Biography. I. Title.
QP511.8.K68A3 1989   574.19′2′0924 [B]   88-32054
ISBN 0-674-30775-5

In memory of Sylvy, my great discovery

Contents

Foreword, by Joshua Lederberg   ix

Preface   xiii

1   The Vitamin Hunters   1

2   Joining the Enzyme Hunters   29

3   Never a Dull Enzyme   59

4   Bless the Little Beasties   88

5   The Synthesis of DNA   121

6   Creating Life in the Test Tube   171

7   Astonishing Machines of Replication   207

8   Frontiers in Replication   240

9   Gene Hunters and the Golden Age   269

10   Reflections on My Life in Science   297

Chronology   320

Bibliography   321

Glossary   324

Index   328

Foreword

Arthur Kornberg has played such a commanding role in the bio-chemistry of the gene that it is impossible to tell its history without including the personality that brought so much of that history about. *For the Love of Enzymes* combines personal memoir with scientific exposition—it is both an autobiography of a great scientist and a biography of the science to which he was devoted. This drama has little of the spice of interpersonal conflict, of any "race for the gold." Arthur Kornberg's rivalry has been with a reluctant Nature who demands ingenuity and perseverance before delivering the real prize, the secrets of how the world and its life are contrived.

My scientific acquaintances are almost evenly divided between those who were born with a passion for science, who have been driven by an inner vocation, and those who came to science as a later discovery, even an accident. Arthur Kornberg belongs to this second category; perhaps that explains the unflagging and methodical way in which he has pursued one accomplishment after another for four decades. The sense of balance and moderation he has brought to his professional life has allowed him to remain deeply devoted to his family, with no noticeable decrement to his scientific productivity. On the contrary, he has enjoyed not just companionship but also laboratory collaboration with a gifted wife (the late Sylvy Kornberg); and he has seen three sons exhibit extraordinary scientific and professional achievement. His talent and sensitivity as an administrator have built a department of biochemistry whose productivity is unmatched, and one which gives the lie to the proposition that science today is achievable only with immense groups and huge machines, or that it demands a renunciation of other human values.

Arthur's early life (like my own) typifies second-generation immigrant Jews in New York City, whose parents made great sacrifices to ensure an education for their children. Nothing in his home background pointed to science except the encouragement to study and to excel. The public schools reinforced that acculturation, by holding forth the ideal that academic achievement was the road of opportunity for social and economic mobility out of the sweatshops. City College in New York has nurtured a lion's pride of Nobel Prize winners, but it offered very little actual science in its undergraduate curriculum (and in those days none whatever at the graduate level). It did offer a talented, ambitious, and competitive peer group that helped to sharpen the aspirations of its students, and a faculty that, whatever else, fed the sense of individual worth of each of them, notwithstanding attributes of race, color, or economic class.

The world outside of City College was not so receptive; there were few opportunities open to Jews in academic or industrial science. The professions such as medicine (however the medical schools might ration their admissions) at least offered a prospect that individual careers might be built on ability rather than on membership in the right groups. A decade later, the mobilization of the universities to train young men and women in the skills needed during World War II finally cracked those ethnic barriers. Indirectly, the same process opened up the National Institute of Health, which gave Kornberg his first research opportunities.

I met Arthur Kornberg 35 years ago, at a summer course given by C. B. van Niel, where Arthur "learned microbiology" in preparation for taking that chair at Washington University. That fateful meeting led, five years later, to my appointment to Stanford University's department of genetics. I would indeed have preferred to be joining Arthur's new biochemistry department; but he and I have differences in how (or whether) we voice a philosophy of science that he may have been wise to foresee, even in our first meeting.

Kornberg's manifest approach to the choice of scientific problems is to focus on the particular, to eschew large social or scientific goals, to set aside grand design and theoretical synthesis. Every detail of enzymology is interesting to him; he says, "I have never met a dull enzyme." Yet, if not by design, then by intuition, Kornberg has always managed to sight the central targets of biological inquiry in his enzyme-hunting, and his method has consistently embraced far broader issues than the mechanical steps of chemical purification and isolation.

Kornberg may be right that enzymology is daunting to many impatient youngsters and, lamentably, is being bypassed in favor of the more facile doctrines of gene hunting. I agree with him that the thousand-odd enzymes involved in intermediary metabolism and in the synthesis of nucleic acids and proteins are the fundamental periodic table of biology. As with the chemical atoms in Mendeleev's time, only a small proportion of the enzymes we can infer have actually been isolated; the arduous and indispensable task of isolating the rest must not be overlooked as the work of mapping the human genome commences in the next decade. But besides Kornberg's technical skills, we will need his knack for selecting those biochemical goals that warrant first priority. As he has always practiced, if not preached, we need to embed the knowledge of enzymes in a broader panoply of their functional relationships in the cell. This will require a host of other skills, like electron microscopy, MRI spectroscopy, and x-ray diffraction, not to mention genetic analysis. But the final test of our analytical methods is to reconstruct the cell, and for that we must have the purified components. Explanation in contemporary biology will continue to be chemistry. No one has taught us that lesson better than Arthur Kornberg.

Joshua Lederberg

Preface

With so much of the attention to DNA and the engineering of genes focused on patent chases and the ethics of genetic manipulations, little is heard about the fascinating origins of this extraordinary revolution in biology. The story of how it came about depends on where the viewer stood when it all happened. Before the mists of history envelop the key facts and figures, I want to give a personal account of how steady advances in biochemistry and studies of enzymes have enabled us to understand heredity and much of life as chemistry. The main impetus for me to tell this story came from the tragedy of my wife Sylvy's long and unremitting illness and the welcome diversion we found in recalling our adventures in science so inextricably woven into our lives together.

Among the many sources of advice and encouragement, I want first to acknowledge the catalytic invitation from the Alfred P. Sloan Foundation to prepare an autobiography. Along the way, it became apparent that my story would not fit into their series because it dwelt more on science than on my life in it; I wished to use the chronology of my career only to organize the narrative and introduce personal elements where they might leaven and humanize the science. I was urged to continue in this vein by my family and many friends, among them Bruce Armbruster, Joshua Lederberg, Robert Lehman, and Wallace Stegner. Later on, I profited greatly from suggestions made by Barbara Bowman and particularly by Gunder Hefta and Susan Wallace of Harvard University Press. To my wife, Charlene, I am deeply indebted for her devotion and inspiration to persevere in this effort, and for the illustrations that form an integral part of the narrative. The contributions of Le Roy Bertsch and Michael Maystead to the graphics, and of Betty Bray to typing the manuscript, were invaluable. The elegance of the book's design I owe to Bob Ishi.

Arthur Kornberg

For the Love of Enzymes

Chapter 1

The Vitamin Hunters

Medical science in the twentieth century has seen a succession of hunters. The bacteriologists, called "microbe hunters" by Paul de Kruif in his popular book published in 1926, occupied the spotlight in the first two decades. They were replaced in the next two by the vitamin hunters. Then the enzyme hunters filled the scene in the 1940s and 1950s, and for the past two decades the gene hunters have been in fashion. Having hardly begun to consume their prey, the gene hunters are already seeing their reign challenged. Should the neurobiologists succeed in developing effective new molecular techniques, these hunters—we might call them "head hunters"—may dominate the last part of our century.

Each age, with its particularly bountiful quarry, was seen as golden. The current age of gene hunting, with its inexhaustible source of genes and a cheap, efficient arsenal to capture them, is incontestably the most golden. There has been no other hunting to match it. Recombinant DNA, the engineering of genes, and related techniques of DNA chemistry constitute what may well be the greatest technological advance in the history of biology and medicine. Genes can now be modified and rearranged at will, and the complete analysis of the human genome is within grasp. Yet, only a few decades ago genetics was the most abstract of the biological sciences, and many people even questioned whether heredity was determined by known physical principles.

How genetics came to be straightforward chemistry can be traced through a chain of discoveries which include my explorations into the enzymes that make the DNA building blocks and assemble them into genes and chromosomes. Unlike the chronicles of polar explorers, the reports of scientific discoveries which appear in professional journals are terse and logical, stripped of personal drama and the true sequence of events. In this account, I will try to give a chronological and simple record of my circuitous journey from medicine to biochemistry and the intricacies of how we make DNA inside and outside ourselves. Mainly, this is a story of science, one which describes many of the discoveries that made the current excitement about DNA, genes, and biotechnology possible.

Growing Up without Science

My degree in medicine and lack of formal training in science surprise people. When they ask whether I actually looked after patients, I tell them, with a touch of pride, that I was a conscientious and able physician, first as a hospital intern and then as a ship's doctor for a crew of five hundred during a wartime stint in the Navy. My career ambition had been to practice clinical medicine, perhaps leavened with some research, until I landed in the National Institute (later Institutes) of Health, where I became an eager investigator of rat nutrition. Three years later I responded to the lure of biochemistry and have remained faithful to it ever since.

I had chosen medicine because I was an avid student and, at age nineteen, with a Bachelor of Science degree from the City College of New York, I welcomed the haven that medical school would provide for four more years. There were no better alternatives in 1937 in the depths of the Great Depression. My early education in grade school and Abraham Lincoln High School in Brooklyn was distinguished only by "skipping" several grades and finishing three years ahead of schedule. I recall nothing inspirational about my courses except the teachers' encouragement to get good marks. When I received a grade of 100 in the New York State Regents examination, my chemistry teacher glowed with pride—it was the first time in over twenty years of teaching that a student of his had gotten a perfect grade. Once when I boasted about this to my wife, Sylvy, she remarked that she too had gotten 100, not just in chemistry but also in algebra and geometry.

Science was unknown in my family and circle of neighborhood friends. In 1947, when I was working in the Biochemistry Department of Washington University in St. Louis under the guidance of Carl and Gerty Cori, Nobel Laureates in medicine that year, Gerty told me that Carl had collected beetles and butterflies in his youth, and then she asked: "Arthur, what did you collect?" "Matchbook covers" was my sheepish response. What else? They were the dominant flora in the Brooklyn streets where I played and in the subways where my father often risked being trampled as he stooped to add one more to my collection.

My parents, before they married, emigrated to New York in 1900 to escape oppression and the bleak prospects for Jews in the small towns of Eastern Europe. Had they stayed in Austrian Galicia (later Poland), they, and likely their children, too, would have been murdered in a German concentration camp. The paternal family name of Queller (also spelled Kweller), of Spanish (Sephardic) origin, had been abandoned by my grandfather. The army draft was a fate no orthodox Jew could contemplate; to escape it, he had taken the name of one Kornberg who had already done his service.

My father, who in Europe had spent most of his days on horseback as a farm manager, was thrust into the sweat shops of the Lower East Side of New York as a sewing machine operator. He used his meager earnings to bring and settle his parents, sister, and other relatives here and, after marrying in 1904, to support his wife and children. When his health failed after nearly thirty years of this brutal labor, he and my mother opened a small hardware and house-furnishing store in the Bath Beach section of Brooklyn. About nine years old then, I helped by waiting on customers and maintaining the inventory. Our kitchen/dining room was behind the store, the bedrooms above it. We constantly teetered on the edge of poverty but somehow managed to stay out of debt.

Except for biblical and Yiddish training, my parents had no access to the formal education they were determined to provide for their children. My father spoke several languages—Yiddish, Hebrew, English, Polish, Russian, and German—and taught himself to write and read English. With all his energies spent on his family, these linguistic abilities found no scholarly outlet. When retinal detachments destroyed most of his vision, and I had left home for medical school, the family side of a daily exchange of cards and letters was undertaken by my mother, who at age 54, with endearing enthusiasm and misspellings, first learned to read and write English.

After high school, I chose the cachet of City College in uptown Manhattan over nearby Brooklyn College, even though commuting from Bath Beach (near Coney Island) meant three hours a day in crowded subways. Competition among a large body of bright and highly motivated students was fierce in all subjects; only one or two A's were given in a class section of thirty. My high school interest in chemistry carried over into college, but prospects for employment in college teaching or industry were dismal. There were no graduate studies or research laboratories at City College then, and so the possibility of doing research hardly surfaced. Seeing little in the way of alternatives, I became one of two hundred premeds, only five of whom would be accepted to medical school in 1937.

Throughout college I worked evenings, weekends, and school holidays as a salesman in men's furnishings stores. This left little time for study or sleep and none for leisure. With my earnings of about $14 a week (a plum salary in those days), a New York State Regents Scholarship of $100 a year, no college tuition to pay, and frugal living, I saved enough to see myself through the first half of medical school at the University of Rochester.

I enjoyed studying to become a doctor. The medical school curriculum was uncrowded, grades were not divulged, personal competition in a class of forty-four students was minimal, and close contact with a fine faculty was encouraged. Among its distinguished members were George W. Corner in anatomy, Wallace O. Fenn in physiology, both of whom I admired, and George H. Whipple in pathology, George P. Berry in bacteriology, and Walter R. Bloor in biochemistry.

Biochemistry seemed rather dull. Description of the constituents of tissues, blood, and urine dominated the field in the United States in the 1930s. The dynamism of macromolecules and of energy exchanges within cells was still unknown, and the mention of enzymes had hardly penetrated my coursework or textbooks. By contrast, the integration of structure and function presented in anatomy and physiology courses was awesome. In pathology and bacteriology, introductions to disease were so gripping as to provoke in me symptoms of several fatal illnesses: the visible chest pulsations of an aortic aneurysm, the quivering muscles in hand and forearm of amyotrophic lateral sclerosis (ALS), the bizarre white blood cells of leukemia, and the cough and sweats of tuberculosis.

One of my aberrations was genuine and persistent—a mild jaundice, noticeable in the slightly yellow or muddy discoloration of the whites of my eyes. The level of bilirubin (a reddish-yellow waste product derived from the routine degradation of hemoglobin) in my blood was well beyond the usual range. These abnormalities, coupled with occasional stomachaches and an intolerance for fatty food, made me, according to current practice, a strong candidate for gall bladder surgery. I avoided it by discovering seven other students with an apparently benign jaundice like mine. I then examined 17 former patients, chosen at random from hospital records, who had recovered from a confirmed attack of acute catarrhal jaundice (now known as infectious hepatitis), 12 patients with assorted severe diseases, and 29 young, healthy people who served as controls. Blood and urine samples and a variety of liver function tests revealed that my seven fellow students were singularly unable to eliminate bilirubin after an intravenous injection, compared with the control group. These findings, described in my first paper, "Latent Liver Disease in Persons Recovered from Catarrhal Jaundice and in Otherwise Normal Medical Students, as Revealed by the Bilirubin Excretion Test," appeared in the May 1942 issue of the *Journal of Clinical Investigation*, then, as now, a selective medical journal.

Why was the incidence of jaundice and an elevated blood bilirubin level seemingly so high among medical students? Probably it was not the incidence that was abnormal—as many as 5 percent of the general population has this aberration—but rather the medical students' inordinate attention to eye discoloration, which they are trained to notice.

Is the disorder a residue of hepatitis, as I implied? Probably not in most cases. More likely, it is an inborn error of bilirubin metabolism. In 1901 the French physician A. Gilbert had first described this familial type of benign jaundice, but neither I nor others around me were aware of this report. The importance of pedigrees in diagnosis of disease was not widely appreciated in the first part of the century. Sir Archibald Garrod's work on heredity of disease was largely forgotten, and there was little interest in genetics among biochemists and physicians or even among botanists and zoologists. By contrast, in the last twenty years, over sixty reports have appeared on a benign form of hereditary jaundice, called Gilbert disease or Gilbert syndrome. Despite this effort, neither the genetic locus nor the enzyme defect is yet known, and diagnosis and significance of my kind of jaundice still remains vague.

Despite this early foray into medical research, I never really considered it for a career. I expected to practice internal medicine, preferably in an academic setting. The idea of spending a significant fraction of my future days in the laboratory had no appeal. Nevertheless, I had hoped to receive one of the fellowships from the medical school which allowed a few outstanding students to spend a year doing research. I was passed over by every department. The students in my class who were awarded fellowships in anatomy and physiology may well have excelled in those courses. The two coveted fellowships in pathology were another matter. Under the aegis of Dean George H. Whipple, the holders were anointed for academic careers. I was not offered a fellowship in pathology, despite (I learned later) ranking first in the course, nor was I awarded fellowships in other departments where it seemed to me I outranked the recipients. In those years, ethnic and religious barriers were formidable, even within the enlightened circle of academic science (see Chapter 10).

The research I did on jaundice grew out of curiosity about my own problem. I collected samples at odd moments and did the analyses on a borrowed bench, late at night and on weekends. My only advisor was a close friend, Arnold V. Wolf, a graduate student in physiology. My only financial support was $100 that William S. McCann, Chairman of the Department of Medicine, gave me to buy bilirubin from Eastman Kodak. Looking back, I realized that I enjoyed collecting these data. During my internship year I kept on collecting bilirubin measurements, and I started setting up to do more analyses in the small sickbay of a Navy ship soon after I joined it in August of 1942. Fortunately, these shipboard studies were cut short by my transfer from sea duty to research at the National Institute of Health.

Rolla Dyer, Director of NIH, had noticed my paper on jaundice in May of 1942, as had the medical corps of the Army and Navy. Desperate over thousands of cases of severe jaundice among recruits inoculated with yellow fever vaccine, a team of medical officers had come to the University of Rochester Medical School in June to get advice. They first sought out Professor Whipple, who had been awarded a Nobel Prize for his studies of blood diseases. Imagine his astonishment when they asked to see Arthur Kornberg, a medical intern, the other "authority" on jaundice at Rochester.

Joining the Vitamin Hunters

A research post at the NIH was a rare assignment at that time. The tiny NIH corps had few openings, and there seemed little reason for Rolla Dyer to select me. Except for the small clinical study on jaundice done in my spare time, I had no formal research experience during medical school or during my internship in internal medicine. It probably mattered that I was touted to Dyer by my friend, classmate, and fellow intern Leon Heppel, who had come to NIH directly on the strength of considerable achievement in research. Heppel had obtained a Ph.D. in physiology at the University of California in San Francisco and had continued research in the Physiology Department at Rochester to earn his way at medical school.

The captain of my ship was not displeased with the prospect of my transfer. He had been repeatedly exasperated by my ignorance of military discipline and indifference to his supreme authority. A heated discussion of some medical administrative matter usually ended with his saying: "I'm the captain." To which I would reply, "But I'm the doctor." That brief duty in the Gulf of Mexico as a ship's medical officer gave me the leisure for some novel opportunities, such as performing the hemitonsillectomy I am about to boast of.

Once when the ship was in St. Petersburg, its home port, another Public Health Service doctor there, who was trained in ear, nose, and throat surgery, offered to show me how to do a tonsillectomy. The operation was still popular then. A recurrent sore throat and enlarged tonsils was considered good reason to perform surgery, and so I had no trouble in finding a willing subject among the ship's crew. With the sailor seated in a chair, my instructor injected a local anesthetic into the area surrounding the left tonsil. Using a snare—a syringe-like instrument with a wire loop at the end—he lassoed the olive-like tonsil. He then pushed the plunger of the syringe and out popped the tonsil, as simple as that. Now it was my turn to do the right side. Apparently when I injected the anesthetic, the needle went in and out of the tissue, delivering most of the drug to the patient's throat and leaving some areas around the tonsil unanesthetized. When I closed the snare, he went straight up out of the chair and let out an ear-splitting shriek. The next day when I saw the sailor on the ship's deck I tried to avoid him, but he followed me to say: "Doc, the side you did feels great but the one that butcher did is killing me."

A year or so later, Dr. Bernard Davis, whom I got to know at NIH, told me about a sailor who had stopped him on a street in New York. Noting on Bernie's uniform the uncommon insignia of a caduceus surrounded by an anchor, the sailor asked: "Are you a doctor in the Public Health Service?" Surprised at this recognition, Bernie asked him how he came by this knowledge. The sailor explained and then asked: "Do you by chance know Dr. Kornberg?" "It so happens I do," Bernie replied. "Great surgeon," said the sailor.

The Nutrition Laboratory at NIH, to which I was assigned in the fall of 1942, had been started by Joseph Goldberger (1874–1929), one of the first scientists to recognize that a vitamin deficiency can cause an epidemic disease. In tracking down the vitamin missing from the diets of pellagra patients, Dr. Goldberger emerged as one of the greatest of the vitamin hunters. The vitamin he discovered was niacin, a member of the B complex of vitamins. W. H. (Henry) Sebrell, whom Goldberger had trained, was now chief of the laboratory and my senior boss. The laboratory had moved in 1938 from downtown Washington to suburban Bethesda, but some of Goldberger's animal caretakers, kitchen staff, and diet notebooks, as well as his aura, were still around.

My work at NIH contributed in a small way to the isolation of another vitamin in the B complex, folic acid. But in spite of this minor achievement, I always felt that I had come to the field of nutrition research in its twilight, decades too late to share the excitement and adventures of the early vitamin hunters who had solved riddles of diseases that had plagued the world for centuries. My envy of their exploits would eventually impel me to search for a new frontier. The discoveries of each of the vitamins became part of my heritage as I went on to learn their biochemical functions. Today, when diet remains as controversial as politics and when the science of nutrition is a shambles, it is important to recall the monumental contributions made by these investigators.

From Microbes to Diet

In the final decades of the last century, one after another affliction had been tracked to a transmissible microbe. It seemed inevitable that similar hunting would succeed with all diseases. There were, however, several common diseases—beriberi, scurvy, and pellagra—for which the causative microbes were proving to be much more elusive than those of anthrax, tuberculosis, and cholera had been.

One hundred years ago, epidemics of beriberi were destroying the Japanese navy. More than half of a crew, after a few weeks at sea, would become weak, listless, and paralyzed and would succumb to profound weight loss, liver disease, and heart failure. Barely twenty-five years after Commodore Perry's visit, Japan had replaced Samurai swordsmen with a navy that would soon challenge one of the mightiest of the West. But unlike Western sailors, the Japanese were peculiarly vulnerable to beriberi despite the best hygiene and the finest rice money could buy; the kernels, having been separated from the ugly husks, were polished free of the protective silvery skins consumed by the population at home.

It struck one K. Takaki, a ship's doctor, that the Japanese had copied every detail of British naval equipment and operations except for rations, and so he designed this crucial experiment: the crew of 300 of one vessel on a long cruise was fed the polished rice diet, while the crew of another was given the unappetizing fare of British seamen: oatmeal, vegetables, fish, meat, and condensed milk. Of those fed rice, two-thirds contracted beriberi; the sailors fed the strange British diet all remained hale and hearty. The dramatic effects of changing to a new ration are preserved in Japanese naval records (see Table 1-1). Although Dr. Takaki could not explain the

TABLE 1-1
Japanese naval records of deaths from beriberi

Year	Diet	Total navy personnel	Deaths from beriberi
1880	Rice diet	4,956	1,725
1881	Rice diet	4,641	1,165
1882	Rice diet	4,769	1,929
1883	Rice diet	5,346	1,236
1884	Change to new diet	5,638	718
1885	New diet	6,918	41
1886	New diet	8,475	3
1887	New diet	9,106	0
1888	New diet	9,184	0

causal relationship between diet and beriberi, to knowledgeable microbiologists in Japan and Europe there was only one interpretation: the polished rice rations must have been infected.

When Christiaan Eijkman, a young Dutch doctor, made his second trip to the Dutch East Indies in 1887 to find the microbe of beriberi, he felt better prepared than for his previous effort, having received bacteriologic training in the celebrated laboratory of Robert Koch. A raging beriberi epidemic in Sumatra and Java was decimating the Dutch troops but sparing the natives they were trying to subjugate. How strange that a microbial infection should afflict only foreigners, despite their advanced hygiene and refined living standards.

A bacillus that Professor Pekelharing, chief of the Dutch commission, had discovered in the blood of beriberi patients did not always produce disease in chickens, nor could Eijkman account for the disease appearing in uninoculated birds. Yet the symptoms in chickens were strikingly similar to those in humans: a nerve involvement resulting in weakness and paralysis of the limbs and a lassitude ending in death. To Eijkman's glory, he took notice of a curious coincidence. With or without inoculation, the chickens stayed well when fed the standard ration of cheap crude rice but succumbed when mistakenly given polished rice reserved for patients.

In well-controlled experiments, Eijkman showed that polished rice produced the beriberi-like disease in chickens and that natural rice cured it. Experiments on prison inmates were just as conclusive. Yet it took many years before beriberi was generally believed to have a nutritional origin and even longer before the true cause was identified.

Illustrative of this long lag in exploiting Eijkman's discovery was the severe outbreak of beriberi in a Manila prison in 1902, some ten years after his report. Social welfare workers who followed in the wake of the United States' occupation of the Philippines were distressed by the unappetizing natural rice fed to prisoners and had it replaced by highly polished kernels glazed white with talcum. Within ten months there were 4,825 cases of beriberi and at least 216 deaths, compared with two cases and no deaths the year before. The American doctors searched intensively for the microbial agent and only belatedly discovered Takaki's report of 1885 about a comparable experience with refined rice fed to the Japanese navy; Eijkman's several publications on beriberi had not even surfaced. Not until 1910 was polished rice replaced by unrefined rice in Philippine military rations, and even as late as 1947 the annual death rate from beriberi was still 24,000 in the population at large.

Why does polished rice cause beriberi? Eijkman attributed the disease to a toxin in rice which is neutralized by the kernel skins lost in polishing. Understanding beriberi as a nutritional deficiency took many years, and identifying the specific lack of vitamin B1 (thiamine) took longer still.

Few present-day physicians in the United States and Europe have ever seen a patient with scurvy or rickets; yet for centuries these two major afflictions killed large numbers of people in the Western world. Like beriberi, scurvy was a scourge primarily of sailors. When Vasco da Gama made his celebrated voyage around the Cape of Good Hope in 1498, one after another of his crew became sick with inflamed jaws, bleeding gums, and loose teeth, followed by internal hemorrhages and anemia. Of the 160 men who began the journey, 100 died of scurvy. In a 1593 report, a British admiral, Sir Richard Hawkins, cited 10,000 deaths from scurvy in sailors that had been under his command, a toll that would have been higher still had he not stumbled on the curative effect of lemons and oranges along the way. As only one of the very many folk remedies, the citrus cure was forgotten until rediscovered in a remarkable way.

The story is told of a sailor dying of scurvy who was put ashore on a desert island to prevent the spread of disease. Gnawing at some grass sprouts, he regained enough strength to hunt for shellfish and snails and was eventually rescued by a passing ship. Learning of this miraculous recovery, James Lind, a British naval doctor, in 1753, treated and cured his scurvy patients with fresh greens and fruits, including lemons and oranges. Yet he was unable to convince the commissary officers of the extraordinary value of these simple supplements to their diet of salted meats and hardtack. It took another hundred years before citrus juices came into regular use in the British navy, whose seamen subsequently became known as "limeys." Why did it take so long? In addition to prejudices, there were some discouraging failures, such as the use of lemon juice boiled to the point of destroying its antiscurvy activity.

Seafarers on long voyages were not the only victims of scurvy epidemics. Severe outbreaks occurred during harsh winters in Europe and among settlers in the New World. Potatoes saved many lives, as did a pine-needle brew provided by friendly Indians. Scurvy had appeared in London in 1883 in the strange guise of

Barlow disease. As Dr. Thomas Barlow described it, fat and rapidly growing babies lost their appetite, bled from their skin and gums, and had painful defects in their bones. They were the victims of a popular, highly caloric, processed food; fortified with minerals and nonperishable, it lacked the unstable antiscurvy substance, vitamin C.

How vitamin C was finally identified as ascorbic acid proved as fascinating as earlier episodes in its history. I heard about the discovery many times from one of the principals, Joseph Svirbely, with whom Leon Heppel and I dined often during my bachelor year at NIH. Joe had done his predoctoral work at the University of Pittsburgh under C. Glen King, whose laboratory was among several devoted to the isolation of the antiscorbutic substance. Their assay used guinea pigs, which, like humans but unlike most animals, cannot make vitamin C from glucose and must rely on their diet for a supply of it.

Hungarian by extraction, Joe obtained a Hungarian exchange fellowship for postdoctoral training in Szeged with the remarkable Albert Szent-Györgyi, who had isolated in crystalline purity a novel carbohydrate that retarded oxidations. Ignorant of its function or precise structure, he had called it "ignose," then "God-nose," and finally, under pressure from the editor of the *Biochemical Journal*, hexuronic acid. He originally obtained it from citrus juice while he was in Gowland Hopkins' laboratory in Cambridge and later from cartloads of adrenal glands during a visit to E. C. Kendall's laboratory at the Mayo Clinic. When Svirbely arrived in Szeged in 1931, he set up the guinea-pig assay as he had in Pittsburgh and tested the adrenal hexuronic acid. One milligram prevented scurvy for eight weeks!

Searching everywhere for a source better than tiny adrenal glands or sugar-rich lemon juice, Szent-Györgyi and Svirbely heeded the suggestion of Mrs. Nellie Szent-Györgyi, inspired one evening by a stuffed-pepper dish. From Hungarian red peppers (paprika), a major crop of the area, they easily obtained kilograms of hexuronic acid, some of which W. N. Haworth in England used to deduce its structure. Szent-Györgyi could now rename his substance ascorbic acid and become the Nobel laureate of 1937 (in medicine or physiology) for a finding which King bitterly contended belonged to him. Caught in the middle of a protracted brouhaha, our ingenuous Joe was unfairly cast as a purveyor of intellectual property (laboratory secrets) when he was really the catalyst of a major discovery.

Less lethal than scurvy, but just as devastating, were the ravages of rachitis, as rickets was called in the seventeenth century. Stunted in growth, with badly bowed legs, a hunched back, and a compressed, pigeon-like chest, stricken children from the sunless alleys of northern Europe were doomed for life. Fish liver oil was believed by many to prevent or control the disease, but its use was irregular because, like other nostrums, its actions seemed rooted in faith and magic. The dietary basis of rickets was also complicated by the apparently unrelated prophylactic and therapeutic value of sunshine. Children in clear and sunny climates were immune. In this century, rachitic youngsters responded dramatically in a few weeks after daily exposures to newly introduced sunlamps.

Until nutrition matured as a science in the first half of this century, the dietary nature of the antirachitic vitamin D, the remarkable potency of cod liver oil, the importance of sunlight, and the skeletal deformities caused by their absence would remain controversial, so that large numbers of children would still be maimed by the disease every year.

Nutrition Becomes a Science

Frederick Gowland Hopkins (1861–1947) is my favorite biochemist, surely one of the greatest of all time. His clear vision of the body economy and vitamins set nutritional science on a path that made the avoidance of deficiency disease the foremost medical advance in this century until the advent of antibiotics.

The upbringing of outstanding scientists commonly includes inspiration by a teacher and the guidance of a successful scientist or laboratory. Hopkins had neither. After an indifferent middle education, he worked as an analytical chemist in toxicology and forensic medicine. Realizing he needed more formal medical schooling, he attended London University as an external (night) student and then Guy's Hospital Medical School. When he finally obtained his medical degree in 1894, some ten years off the normal course, he still had no formal training in either biology or chemistry. What he did have in his favor was a deep curiosity about the makeup and operations of living things, the discipline to rigorously apply chemistry to isolate and identify the molecules responsible for them, and a sound training in clinical medicine.

One of Hopkins' lifelong research interests was the chemical nature of the pigment of butterfly wings, an interest accompanied by a deep faith that understanding these pterin pigments, as he called them, would have broad significance and ultimately some clinical relevance. Knowledge of the pterins did in fact prove to be crucial in figuring out the structure and functions of folic acid and, later on, in the effective use of folic acid antagonists in the chemotherapy of cancer.

In a lecture to students near the turn of the century, the conventional picture of human and animal nutrition that Hopkins could paint was rather simplistic: The body needed fuel to produce the energy for its work. As with a combustion engine, what mattered was the energy content of the fuel rather than its particular chemical composition. Thus it seemed that carbohydrates, fats, and proteins could be used interchangeably. Vegetables, having a low caloric content, were frivolous as a food source. Proteins were regarded by some as having special value, but here again no major nutritive distinction was made as to their source.

But Hopkins' own research and that of others would soon revise this picture. The striking effects of rice polishings in beriberi, of fresh greens and citrus juice in scurvy, and of fish liver oil in rickets were beginning to implicate special factors in the diet unrelated to caloric value. And back in 1881, scientists in Basel had been surprised to find that mice given a diet of the known components of milk—lactose (a sugar), casein (a protein), fat, and minerals—could not survive, while mice fed whole milk behaved normally. They concluded that "a natural food such as milk must therefore contain besides these known principal ingredients small quantities of unknown substances essential to life." However, these conclusions were not pursued to determine the identity of the substances missing from the "dissected milk," nor was there a followup of similar conclusions reached by a Dutch scientist twenty years later.

To Hopkins we owe the first systematic and thorough attempt to formulate a synthetic or purified diet in order to identify all the factors needed to sustain the growth and health of an animal. With rats, which were to become the classic subjects for nutritional studies, he learned that casein, "if thoroughly washed with water and alcohol lost its power of support, while addition of the very small amount of the washings (of the casein) restored this power"; extracts of yeast proved to be even more potent in supplementing the washed casein. Hopkins realized from these studies that "accessory growth factors" were essential dietary constituents. In 1912 these factors were named *vitamines* by Casimir Funk: "vita" for life, and "amines" because they were mistakenly thought to have an amine chemical structure.

At one point in Hopkins' experiments, a previously adequate synthetic diet failed repeatedly to sustain his rats, until he realized that he had substituted lard for butter as the fat supply. Butter, but not lard, is a source of the fat-soluble vitamins A and D, discovered later by Elmer V. McCollum (1879–1967), the distinguished American nutritionist.

Just as Joseph Goldberger had in his career linked the microbe hunting of the previous century to the vitamin hunting of this one, so did Hopkins bridge the vitamin era and modern biochemistry. At that time, cellular ingredients were lumped into an amorphous mass, called protoplasm, and a vitalistic view prevailed in which cellular operations were regarded as comprehensible only in the intact living cell. Hopkins was the bold knight who fought the obfuscations of both protoplasm and vitalism. Blending biology with chemistry, Hopkins sought the molecular basis of how cells and organisms manage their lives. He articulated the philosophy of biochemistry better than anyone before or since.

Beyond his experimental discoveries and philosophic insights, Hopkins started and led the Cambridge group, the most productive school of biochemistry for two decades. Kind and diffident, generous but firm, he was adored and revered as "Hoppy" by disciples from diverse national origins, including many refugees to whom he gave haven and whom he launched on illustrious careers.

Dermatitis, Diarrhea, and Dementia

Joseph Goldberger became one of my heroes for more than the sentimental associations of my being his scientific "grandson" and sharing some of his cultural antecedents. In 1914, when Goldberger was sent to the southern United States to find the microorganism responsible for a major outbreak of pellagra, the spotlight of medical science had not yet shifted from the illustrious microbe hunters Louis Pasteur and Robert Koch to the emerging vitamin hunters. Through his discovery on this assignment, Goldberger became a key figure in this transition.

Joseph Goldberger emigrated to the United States from Europe in 1881 at the age of seven and grew up in poverty in New York's Lower East Side. Yet he managed to go to City College of New York and later to get medical training at Bellevue Hospital Medical College, now New York University Medical School. He joined the U. S. Public Health Service as a commissioned officer seeking adventure, and he soon found it in the Hygienic Laboratory (renamed the National Institute of Health in 1930). The Laboratory focused heavily on infectious diseases. Goldberger pursued and made major contributions to the understanding and control of several of them, including yellow fever and dengue fever, each of which nearly killed him.

Many before him had sought a microbe responsible for pellagra, without success. Named, in Italy, "pelle agra," for the rough skin lesions it caused, the disease also resulted in severe weakness, intestinal disturbances, and nervous system disorders which often led to commitment to insane asylums (as mental hospitals were then called). Each year, epidemics of pellagra affected hundreds of thousands of people in the southern states. Economic repercussions were severe on the cotton plantations where many of the victims were employed. Goldberger ingested and inoculated himself with skin scales, nasal mucous, and excreta of patients in order to locate the causative microbe, but he did not contract the disease. Nor did any of his collaborators, who, encouraged by his heroic acts, also volunteered for such trials. When Goldberger visited institutions having severe epidemics, he was struck by a remarkable oddity for a contagious disease. While the inmates were affected, the nurses, doctors, and administrative staff were not.

Goldberger astutely noticed a difference in the diets of inmates and staff. The inmates were fed corn bread, grits, molasses, and fatback (lard); the staff ate lean meat, milk, and vegetables. When he returned to his laboratory in Washington and fed the inmates' diet to dogs, they developed blacktongue, the pellagra-like state he had seen in the sorry mongrels of poor rural areas. When he fed the dogs meat, milk, or vegetables, they were promptly cured.

In a carefully controlled experiment, eleven prisoners were fed a calorically adequate diet containing corn flour, refined wheat flour, potato flour, salt pork, and syrup; in six months seven of them came down with pellagra. They recovered rapidly with a daily ration of 30 grams of yeast, 200 grams of lean beef, and a quart of milk. Now thousands of pellagra patients in hospitals could also be miraculously cured by this simple dietary regimen, and many previously considered hopelessly insane could be discharged from the asylums.

For some ten years Goldberger tried to isolate the substance in yeast, liver, and other foods responsible for the prevention and cure of blacktongue and pellagra. After his death in 1929, another eight years passed before D. Wayne Woolley and Conrad Elvehjem at the University of Wisconsin finally succeeded. To their disappointment, this member of the vitamin B complex was not a novel (patentable) substance but rather a well-known chemical, nicotinic acid or niacin, an oxidation product of nicotine, widely distributed in nature and the active principle of tobacco. Ironically, what had inspired Casimir Funk twenty-five years earlier to recognize the existence of vitamins was the isolation of a substance which he mistakenly thought cured beriberi but was in fact the antipellagra vitamin nicotinic acid.

Anemia, Sulfa Drugs, and Folic Acid

My initial project as a nutritionist at NIH was to find out why rats fed a purified ("synthetic") diet (consisting of glucose, 73%; vitamin-free casein, 18%; cod liver oil, 2%; cottonseed oil, 3%; salts, 4%; and the vitamins thiamine, riboflavin, pyridoxine, niacin, and pantothenate [and choline]) remained sleek and grew normally, but when certain sulfa drugs (sulfadiazine or sulfathiazole) were added, more than two-thirds of the rats developed a severe blood disorder in a few weeks and died. The two major diseases that occurred were anemia (decrease in red blood cells) and granulocytopenia (decrease in granulocytes). A stock animal ration or inclusion of a liver supplement in the purified diet was effective in preventing and curing these sulfa-drug diseases. It seemed therefore that we were dealing with an induced dietary deficiency rather than a drug toxicity.

My colleagues and I in the Nutrition Laboratory were aware that other vitamin hunters were following the scent of a new factor which might be the one lacking in our purified diets. Monkeys and chickens fed such diets, supplemented with all the known vitamins, developed anemia. A factor in yeast, liver, and vegetables assayed for its prevention of anemia in monkeys had been called vitamin M; an antianemia activity from similar sources assayed in chickens was named vitamin B_c by some and factor U by others. There was also an anemia associated with the diarrheal disease tropical sprue, as well as with other human anemias in infancy and pregnancy; all were responsive to dietary supplements, implicating a factor possibly related to an earlier reported "Wills factor." (Lucy Wills, an English physician, observed in 1931 that an anemia in pregnant women in India, consuming the common high-wheat diet, could be cured by yeast or liver extracts.) As for the anemia in our rats, it seemed plausible to us that the intestinal bacteria normally synthesize this antianemia vitamin and supply it to their host, and that the sulfa drugs, by destroying the gut bacteria through their antibiotic action, removed the source of this essential factor.

A powerful new weapon in vitamin hunting had been introduced about this time: microbes as "experimental animals." Using rats to assay the potency of a factor requires many animals, much labor and expense, and many weeks of waiting for the deficiency disease to develop. The difficulties are progressively greater when the assay animal is a chicken, monkey, or human. With bacterial populations of hundreds of millions doubling every half hour in test tubes, measurement of their growth by the turbidity or acidity produced in the medium is precise and reliable. The cost and labor are trivial, and the results are in hand within a few hours. Moreover, all concerns with animal welfare are avoided.

The nutrition of most bacterial species is relatively simple. For example, *Escherichia coli* (*E. coli*), the most prominent inhabitant of the intestinal tract, grows nicely in a medium composed only of glucose and salts. Each bacterial cell can make all the protein building blocks (amino acids) and vitamins it needs. But occasionally a mutant of the species occurs naturally or is produced in the laboratory which has a defective gene and thus lacks the equipment to produce a given vitamin, say thiamine. If this mutant is to grow, thiamine must be present in the medium. The growth rate of this mutant, called *E. coli* thiamine$^-$, is an accurate assay of the presence or absence of thiamine.

In surveys of thousands of naturally occurring species of bacteria, a few species were found which required a factor in yeast, liver, or vegetables that was not replaced by any of the known vitamins. These naturally defective bacteria provided the key for those who

Folic Acid

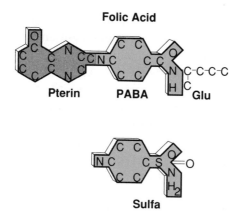

FIGURE 1-1
Sulfa drugs mimic the PABA constituent of folic acid,
thereby preventing bacteria from synthesizing the vitamin
and thus inhibiting their growth.

sought simple assays for essential missing nutrients. One of these
exacting bacterial species, *Lactobacillus casei* (*L. casei*), is prominent in cheese fermentations. This organism cannot grow in a simple
culture medium that does not contain liver or yeast extracts and
therefore became the assay standard for purification of a factor from
liver and yeast that proved to be the same as the antianemia factor.
This work was done by E. L. R. Stokstad at the Lederle Laboratories
in Pearl River, New Jersey. Similarly, the growth of the bacterium
Streptococcus lacti (*faecalis* strain) guided the isolation by Herschel
Mitchell (then at the University of Texas at Austin) of what turned
out to be an identical substance from spinach. Two tons of spinach
yielded ten milligrams (three ten-thousandths of an ounce) of a substance with acidic properties; in recognition of its leafy source, it
was named folic acid (Latin *folium*, leaf).

Daily doses of 25 micrograms (a microgram is one millionth of a
gram; 28 grams make an ounce) of *L. casei* growth factor cured rats
severely deficient in white blood cells in four days. Anemic chickens, monkeys, and people responded dramatically to the folic acid
isolated from animal or plant sources. When the structure of folic
acid was solved, it could be seen as a molecule with three constituents (Fig. 1-1). One was a pterin, closely related to the pigment
F. G. Hopkins had discovered in butterfly wings. Another was para-
aminobenzoic acid (PABA), an entity which sulfa drugs resemble

and mimic; it is the same PABA molecule used in sun-block lotions to absorb UV light. The third constituent, glutamic acid, is one of the amino acids in proteins. (The common name for this vitamin, folic acid, has survived serious attempts to replace it with the chemically descriptive "pteroylglutamic acid" or even the simple "folacin.")

With the isolation of folic acid, it was clear to me that virtually all the vitamins had been discovered. But we still did not understand what most of the vitamins actually do in the body. How does folic acid serve in the growth of blood cells? What clues does the structure of folic acid offer to understanding its precise metabolic function? Can this understanding explain why sulfa drugs kill bacteria but not animal cells and why folic acid analogs are potent agents in the chemotherapy of cancer? The answers to these questions, as well as similar questions about the functions of the other vitamins, would be answered in the next two decades by the enzyme hunters. Just as the microbe hunters, who held sway in the first two decades of this century, were succeeded in the 1920s and 30s by the vitamin hunters, so would the vitamin hunters be overrun by the enzyme hunters, who would in their turn give ground to the gene hunters of our present era.

Intestinal Bacteria, Blood Clotting, and Rat Poison

Because sulfa drugs caused anemias and white blood cell deficiencies only when rats were fed highly purified (synthetic) diets, it seemed clear to me that these drugs produced these diseases by destroying the intestinal bacteria on which many animals rely for an adequate supply of folic acid and other vitamins. I was therefore puzzled by a scientific report which appeared in 1944 casting serious doubt on this interpretation. In pursuing this question, I became familiar with the thread that ties intestinal bacteria to blood coagulation and eventually to the best rat poison of all time. The thread is vitamin K. The story successively involves the nutrition of chickens and rats, a disease of cattle caused by consuming spoiled sweet clover hay, the intricate chemistry and clinical treatment of blood clotting, and even the formation of bones and coral reefs.

In 1929 Henrik Dam, a Danish scientist, observed severe spontaneous hemorrhages in the chickens he was feeding synthetic diets;

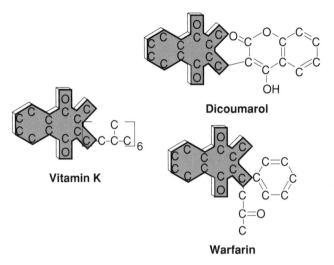

FIGURE 1-2
Dicoumarol and Warfarin, structural analogs of vitamin K,
interfere with its blood-clotting functions.

their blood took ten times longer to clot. Because various green leafy
materials and dehydrated alfalfa meal prevented and cured the disease, he inferred the action of a new vitamin and called it K for
"koagulation." The pure substance proved to be a naphthoquinone,
a ringed arrangement of carbon atoms with two oxygens and a long
carbon tail (Fig. 1-2).

Chickens proved to be unique in their ready susceptibility to vitamin K deficiency when on a synthetic diet. Rats, by contrast, usually
showed no disorder of blood coagulation until sulfa drugs were
added to their synthetic rations. The intestinal bacteria presumed to
be responsible for producing vitamin K were thought to be too few in
number in the limited space of the chicken's large intestine to supply enough of the nutrient, but were in sufficient abundance in the
voluminous cecum (large bowel) of the rat as long as sulfa drugs
were not added to their diet. (In fact, it was essential that our rats be
kept in separate cages with wire mesh bottoms to prevent them from
obtaining vitamin K by eating their own or others' feces.)

The reported observation that I found difficult to explain at the
time was as follows. Para-aminobenzoic acid (PABA) is a building
block of folic acid which, as we have seen, is very similar to sulfa
drugs (Fig. 1-1). When a sulfa drug enters the bacterial cell, it com-

petes with PABA for the enzyme that assembles folic acid, thereby preventing the manufacture of folic acid and so depriving the bacterium of this life-essential substance. (Animal cells, on the other hand, cannot make folic acid, but animals obtain adequate amounts of the already-assembled vitamin in their diet. The sulfa drug, having no target for its action, proves harmless.)

The report stated that PABA prevented the sulfa drug from producing a blood-clotting defect due to vitamin K deficiency in rats *even when given by injection.* This result was taken to mean that the sulfa drug was not exerting its toxic effect on the animal's intestinal bacteria but somewhere else in the body. Might this interpretation also apply to my work on folic acid, from which I had concluded that the deficiency produced by sulfa drugs was due to their inhibitory action on the intestinal bacteria? Alternatively, I wondered whether injected PABA might find its way through the bloodstream into the rat's large intestine to achieve there a concentration sufficient to overcome the sulfa drug action on the bacteria.

I repeated the experiments with PABA injections and sulfa drugs and developed a method to measure the amounts of vitamin K and PABA in the intestinal contents and feces of rats. It was based on the capacity of a sample (compared with pure vitamin K) to correct the prolonged clotting time of a vitamin K-deficient rat. Recovery (in 24 hours) to levels of 30, 50, and 90 percent of normal clotting ability was obtained when deficient animals were given 2, 3, and 5 micrograms, respectively, of the pure vitamin (Fig. 1-3). But when feces from rats on the sulfa-drug diet were administered (by stomach tube) to vitamin K-deficient animals, recovery did not occur, because the feces contained less than 2 micrograms of vitamin K over a 5-day period, compared with more than 50 micrograms over the same period in the feces of rats fed only the purified diet. Thus, ample quantities of vitamin K were being produced by the intestinal bacteria of rats on a purified diet, and this production was eliminated by sulfa drugs. As for rats *injected* with PABA, just as I had suspected, high levels of this substance accumulated in the intestine, amounts sufficient to offset the action of the sulfa drug taken in the diet. These findings were my first contribution to the prestigious *Journal of Biological Chemistry,* one of the very few biochemical papers emanating from NIH at that time. (Twenty years later, NIH had grown to become the world's most prolific center of biochemistry.)

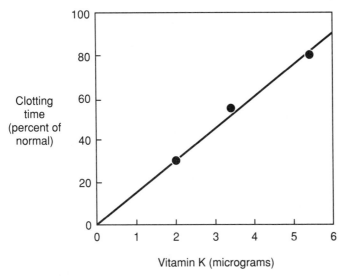

FIGURE 1-3
The vitamin K content of a sample was assayed by restoration of blood clotting a day after its administration to a rat with a vitamin K deficiency.

Meanwhile, literally back at the ranch, the cause of a fatal hemorrhagic disease in cattle was being tracked down to the consumption of spoiled sweet clover hay. The responsible poison was discovered to be a small molecule, of a class called dicoumarols, which strikingly resemble vitamin K (Fig. 1-2). Dicoumarol, in mimicry of vitamin K, was correctly assumed to be interfering with an essential, but then unknown, function of the vitamin in blood clotting. Pursuit of how vitamin K functioned in coagulation took more than thirty years, and dicoumarol played an important part in solving this question.

The coagulation of blood must be finely controlled. A myriad of tiny breaks in the vascular system must be plugged to prevent oozing and hemorrhages. Yet the clotting response needs to be tuned to the severity of the break and the time needed to seal it; excessive clotting leads to the thromboses that cause the coronary occlusions in heart attacks and brain vessel occlusions in strokes. To achieve this fine regulation of blood clotting, a cascade of six or more chemical reactions precedes the final conversion of soluble fibrinogen molecules

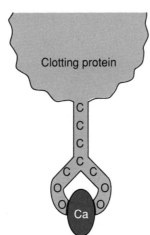

FIGURE 1-4
A protein with a claw-like
appendage orients calcium
for its essential role in
the blood-clotting process.

into the firm fibrin meshwork of the clot. (Hemophilia is an example
of a hereditary deficiency of one of the proteins responsible for a
stage midway in the cascade.) Calcium is an essential component in
several of the clotting reactions, and the responsible proteins are
furnished with special claws to position the calcium atom for its
action (Fig. 1-4). These claws, which despite intensive study for
many decades had escaped detection until recently, are added to the
protein through the action of vitamin K!

Dicoumarols, by interfering with vitamin K functions, cause the
gradual accumulation of clawless clotting proteins that are defective
in binding calcium. This action of dicoumarol has been put to frequent use in the long-term drug treatment of excessive thromboses in
cardiovascular disease. Dicoumarol is related to the ingredient of
the most commonly used rat poison, warfarin (Fig. 1-2), marketed by
d-CON. A major advantage of this poison is its relative safety. A
single ingestion of dicoumarol by a child or pet does not cause bleeding; repeated doses, in a formulation attractive to rodents, are required to prevent vitamin K from replacing the clawed proteins lost
in the wear and tear of the body's clotting activities.

Is the action of vitamin K limited to attaching calcium-binding
claws to blood-clotting proteins? It seems not. Such appendages
have recently been found on proteins crucial in the binding and
deposition of calcium in bones, teeth, egg shells, and coral. Future
research will likely show that vitamin K is responsible for putting
calcium-binding appendages on these proteins as well.

The Decline of Nutritional Science

Books about nutrition, like those about sex and money, are best-sellers because eating is so fundamental to survival, health, and pleasure. People want to know what and how much to eat. Very few people realize how few facts about nutrition have actually been established, and most people have no reliable way to separate these few facts from the faith healing and quackery that masquerade as science. Turning to physicians is of little help because they don't know much about nutrition either, and one cannot teach what one doesn't know. No wonder that people will grasp at books purporting to reveal how to achieve health and longevity by adherence to a few simple dietary commandments. What can be done to advance nutritional knowledge and preserve nutrition as a science?

In 1938 the prescribed medical school course in nutrition I took at the University of Rochester was called Vital Economics. From texts and laboratory exercises, I learned three key precepts: (1) The body uses a certain number of calories at rest or at work that must be matched by consumption of food of sufficient caloric content. Calories are furnished by combustion of carbohydrates, fats, and proteins. Insufficient intake results in loss of weight; excessive intake results in weight gain. (2) Of the twenty building blocks (amino acids) in proteins, the human body can manufacture only ten and must rely on proteins in food to supply the others; some proteins fail to supply all the amino acids needed by the body. (3) Accessory factors (vitamins) and minerals (such as salts of sodium, potassium, iron, calcium, magnesium, phosphorus, sulfur) must also be consumed.

At the level of diet and body economy, relatively little has been added since the 1930s to this core of nutritional knowledge. There were some flickers of excitement left for the vitamin hunters in the next few years, as in the discoveries of folic acid and vitamin B12, but the major challenges in nutritional science lay elsewhere, in biochemistry.

Fashions in science are as influential and nearly as mercurial as styles in dress. Driven by the funding tastes of government agencies and major foundations, the stampede of scientists around the world to fashionable scientific activities leaves ghost towns in still fertile areas. Such was the fate of nutrition as a major scientific discipline when in the 1940s and 50s biochemists were lured away by access to the molecular details of cellular operations. The word "nutrition" no longer appears on the lintel of the expanded biochemistry building at the University of Wisconsin, once the seat of world leadership in a proud science.

The depopulation of nutritional science has not been entirely whimsical. Compared with the classical deficiency diseases, the residual problems of animal and human nutrition are elusive and their scientific pursuit far more difficult, time-consuming, and unglamorous. Take, for example, the consumption of sucrose, which increased many-fold in Western societies at the same time that there was a comparable increase in the incidence of heart disease. Were they causally related? A controlled prospective (rather than retrospective) study of the influence of a dietary ingredient, be it sucrose or cholesterol, on a disease that develops over decades is exceedingly difficult, especially when many factors, such as heredity, exercise, infections, and caloric intake, may contribute in a significant way. Similar difficulties face those who would try to determine the undoubtedly important effects of diet in the prevention and treatment of cancer, immune disorders, and aging. These complicated and perplexing processes are hardly the best neighborhood for the growth of a struggling scientific discipline.

Advances in technology have also complicated human nutrition by giving rise to questions that were previously beyond reach. One advance is the abundant and cheap supply of pure vitamins that has made megavitamin dosage a feasible and common practice. Does a vitamin intake one hundred times that needed to maintain normal levels provide some benefit to a small percentage of the population? Are there circumstances in which such megavitamin dosage may also have adverse effects?

Another scientific advance has given us an exceedingly sensitive test to determine if a food can cause mutations and cancer. A large number of natural foods and food processes are mutagenic and can be presumed to be carcinogenic and also to accelerate aging. Among these are mushrooms, celery, chocolate, bruised potatoes, alfalfa sprouts, black pepper, peanut butter, some herb teas, charred meat, and a hundred other common foods (not to mention those with added mutagenic preservatives and dyes). To monitor an expanding number of incriminated foods, to assess the significance of mutagenic levels, and to formulate a varied, palatable, balanced diet for different age groups in assorted states of health and disease exceeds the limits of available resources and knowledge.

Perhaps the most perplexing problem human nutrition faces is the confusion between nutrition as a natural science and nutrition as the art of feeding people. It is a confusion analogous to that between medical science and health care. There is no question that massive famine in Africa is intolerable, as it is in pockets of urban and rural America. Ten thousand children go blind in India each year from a deficiency of vitamin A which could be prevented by five cents' worth of the synthetic vitamin per child. All ten essential vitamins and eight minerals can be supplied at a bulk cost of less than a dollar per person per year. Widespread deficiency diseases are a failure for which society, not the nutritional scientist, is to blame.

In overreaching its scientific domain, nutrition, like so many other human endeavors, becomes prey to prejudice, chicanery, and other perils of the culture. In trying to encompass medicine, agriculture, economics, psychology, and anthropology, as well as chemistry and biology, nutrition becomes more a social and political activity and less a science.

The disagreement among scientists about the limits of nutritional science was made apparent to me in responses to the "Neuberger Report on Food and Nutrition Research" issued jointly by the British Agricultural and Medical Research Councils in 1975. The report emphasized that "human nutrition is mainly concerned with defining the optimum amounts of the constituents of food necessary to achieve or maintain health." The report recommended more research in the biochemical aspects of nutrition, but it also deplored the relative neglect of human nutrition and urged more clinical and epidemiological studies. Despite these laudable statements, the report was attacked by other nutrition scientists for inadequate concern with economics, anthropology, sociology, demography, and psychology. "Nutrition is the one science that can least afford to remain in the laboratory; it concerns every single human being every single day of his life," said John Yudkin. T. B. Morgan was critical of the report because it failed to stress "the inseparable interrelationships between agriculture and nutrition" and neglected the importance of training graduates from food science, nutrition, catering, and dietetics for research in *social* nutrition.

Whither human nutrition? How can we cope with the numerous and complex problems and the extraordinary difficulty of doing controlled, long-term dietary experiments on humans? Considering the importance of acquiring definitive dietary information for health and economic welfare, nutrition cannot be left to the art of medicine or the exploitation of hustlers. The only answer is science, hard science. Progress demands these actions: (1) invest in the training and support of scientists to work in nutrition, (2) narrow the focus of experimental work to doable problems, urgent or not, (3) use a variety of animal models, (4) insulate research from broad social issues, and (5) sustain the faith that persistent scientific effort eventually solves most problems, often in a surprisingly novel way.

Chapter *2*

Joining the Enzyme Hunters

By 1945 I had become bored feeding rats variations of purified diets and trying to find in their survival, growth, fertility, and abundance of blood cells evidence that still one more vitamin remained to be discovered. In thinking about exactly how folic acid serves the needs of growing cells—such as blood cells and the cells that line the intestines—I was intrigued by a new breed of hunters tracking down the metabolic enzymes, the intricate machines that used vitamins to catalyze the combustion of sugar in yeast and muscle to generate the energy for their growth and work.

These enzyme hunters showed that the yeast cell modulates the combustion of a sugar molecule to extract energy, which it needs for growth and reproduction, in an astonishingly complex series of chemical operations, each catalyzed by a distinctive enzyme. A device of this chemical subtlety is needed to capture and store the energy derived from the combustion of food and avoid wasting it as heat energy. These revelations of how yeast metabolizes sugar provided the methods and direction for the later discovery that muscle also derives energy from sugar to do its work, using enzymes virtually identical to those of yeast.

With these metabolic pathways and enzymes defined, it became clear that the vitamins function in some of the key steps: one enzyme depends on thiamine, another on niacin, still another on riboflavin.

The vitamins serve as cocatalysts or *coenzymes*—detachable working parts of an enzyme—to perform detailed chemical operations. I vividly recall my excitement in reading for the first time about these enzymes, coenzymes, and the miracle molecule called ATP in papers by Otto Warburg, Otto Meyerhof, Carl Cori, Herman Kalckar, and Fritz Lipmann. From my medical school training I had learned nothing about these people or the things they were investigating. I also recall my astonishment on hearing a seminar at NIH by Edward Tatum in which he described the work he and George Beadle had done with mutants of the bread mold *Neurospora*, demonstrating that one gene is the blueprint for one enzyme. I knew even less about genetics than about biochemistry.

As soon as the war was over, I persuaded Dr. Sebrell to let me quit my nutritional work and go to a laboratory where I could learn about ATP and enzymes. While still in uniform, I spent a year with Severo Ochoa at New York University Medical School and then six months with Carl and Gerty Cori at Washington University Medical School in St. Louis. It was then that I fell in love with enzymes.

Although enzymes were recognized in the nineteenth century as catalysts for certain chemical events in nature, their importance was not fully appreciated until their role in alcohol fermentation and muscle metabolism was defined. Then it became clear that virtually all reactions in an organism depend on the high catalytic potency of a cast of thousands of enzymes, each designed to direct a specific chemical operation. Deficiency of a single enzyme—as the result of mutation, for example—could spell disaster for the cellular or human victim. I abandoned the animal nutrition laboratory when I realized that enzymes are the vital force in biology, the sites of vitamin actions, and the means for a better understanding of life as chemistry.

The decades of the forties and fifties became the heyday of the enzyme hunters, who along the way defined the detailed chemical functions of nearly all the vitamins. The vitamin-hunting days were over, and gone with them was the interest of leading biochemists in the influence of nutrition on the whole organism.

The enzyme hunters could have preceded both the microbe and vitamin hunters but for accidents of history. The prize of one kilogram of gold offered by the French Academy of Science in 1779 to anyone who could explain the nature of alcohol fermentation might have been earned by any of the great chemists of the early nineteenth century—Jöns Jakob Berzelius (1779–1848) of Sweden or Justus von Liebig (1803–1873) and Friedrich Wöhler (1800–1882), both of Germany—but for their poor sense of biology and their failure to recognize the central role of yeast in fermentation. Louis Pasteur (1822–1895), an equally great chemist, did appreciate that role, but his exaggerated interest in biology eventually caused him to neglect his chemical roots and thus delayed the advent of modern biochemistry and the discovery that enzymes are the vital force of fermentation. The claim to the French prize would not be made for one hundred and twenty years.

Yeast and the Origins of Biochemistry

Periodic revival of a languishing interest in historic figures and events also applies to subjects in science. The yeast cell, responsible for the birth of modern biochemistry and the source of the very word "enzyme," was ignored for a long time but is now budding again. With its rich legacy of genetic and biochemical data, yeast appears to be the most accessible object for current studies of the nucleus, mitochondria, and other complex features of eukaryotic cells (those, such as animal cells, which have a nucleus). I will therefore take a brief excursion to examine the heroic stature of yeast and show why it is so basic to everything we know and do in biochemistry.

How yeast converts sugar or starch into wine or beer, how it gives champagne its sparkle and bread its leavening, have led biochemists toward an understanding of the molecular basis of cellular behavior. The sweet juice of the crushed grape is transformed quickly into an intoxicating, tasteful wine. Grains steeped in water begin to malt, and soon the fluid is a flavorful, nourishing beer. These fermentation processes were worked for over 6,000 years, but their chemistry remained a total mystery until the nineteenth century. The role of yeast cells in wine making and beer brewing then became the subject of one of the most protracted and vitriolic polemics in biology.

By the time of Antoine Lavoisier (1743–1794), the principal carbon compound of grape juice was known to be sucrose (beet or cane sugar), and the main carbon products of its fermentation were known to be ethanol (alcohol) and carbon dioxide. A French scientist of da Vincian scope, Lavoisier made two major contributions to

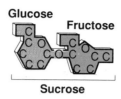

FIGURE 2-1
Sucrose (cane or beet sugar) is made up
of glucose and fructose.

the understanding of fermentation before losing his head to the Revolution. First, he showed that a solution of the sugar itself could be
fermented if provided with the sediment (with negligible carbon
content) from a previous fermentation. Then he demonstrated, by
combustion of samples in air, that the weight of carbon appearing in
the sugar that disappeared in a fermentation was matched by the
weight of carbon in the ethanol and carbon dioxide products. Subsequent improvements in analysis of the elemental composition of
substances allowed a proper balance sheet for fermentation to be
written. With the simpler sugar glucose in place of sucrose (Fig. 2-1),
the elements in each of the participants in fermentation were shown
to be:

Substance	Carbon	Hydrogen	Oxygen	Formula
Glucose	6	12	6	$C_6H_{12}O_6$
Ethanol	2	6	1	C_2H_5OH
Carbon dioxide	1	0	2	CO_2

The overall equation of this fermentation is:

$$\text{glucose} = 2\ \text{ethanol} + 2\ \text{carbon dioxide}$$
$$(C_6H_{12}O_6) = 2\ (C_2H_5OH) + 2\ (CO_2)$$

Just how is a glucose molecule converted into two molecules each
of ethanol and carbon dioxide? Left as a sugar solution, even if subjected to heat, pressure, or acid, this would never happen. Yet these
sugar molecules in grape juice at room temperature are all converted
to the same products in a few days.

Ever since Anton van Leeuwenhoek (1632–1723) saw yeast cells in fermentation sediments, many believed them to be the driving force of the process, particularly Theodor Schwann (1810–1882), to whom we also owe the proposition that cells are the basic organizational units of all organisms. He showed that living yeast cells, often carried in by the air, must be present during a fermentation. His work, along with that of others, also demonstrated that the conversion of sugar by the yeast cells sustained their growth and multiplication. Against this view were ranged the leading chemists of that period. Berzelius had the most prescient ideas about catalysis of chemical reactions but was contemptuous of attributing the chemistry of fermentation to yeast cells. Wöhler had done the dramatic experiment of synthesizing urea from carbon dioxide and ammonia. Up to then, urea had been made only by living creatures. Having disposed of the notion that "vital forces" were needed to synthesize an animal substance from simple molecules, he now ridiculed the proposal that a yeast globule is the chemical factory of fermentation. He teamed up with Liebig in publishing obscene, anthropomorphic caricatures of a yeast cell ingesting and digesting sugar and excreting alcohol and gas.

Against this background, Pasteur did his epochal work confirming and extending the earlier experiments of Schwann on the preeminence of the living yeast cell. Unlike Schwann, Pasteur was effective in countering the prevailing view of the chemists, so much so that he stifled the truth in some of their arguments. Said Pasteur:

> In introducing yeast into a sugar solution we are sowing a multitude of minute living cells, representing innumerable centers of life, each capable of vegetating with extraordinary rapidity in a medium adapted to their nutrition.
>
> The globules of yeast are true living cells, and may be considered to have as the physiological function correlative with their life the transformation of sugar, somewhat as the cells of the mammary glands transform the elements of the blood into the various constituents of milk in connection with their life functions.
>
> My present and most fixed opinion regarding the nature of alcoholic fermentation is this: I believe that there is never any alcoholic fermentation without there being simultaneously the organization, development, and multiplication of the globules, or at least the pursued, continued life of globules that are already present. The totality of the results in this article seem to me to be in complete opposition to the opinions of Liebig and Wöhler.

In the thirty-year war of words with his enemies in chemistry, Pasteur emerged the clear victor but at an intolerable cost. He allied himself with the philosophy of vitalism, which held that life processes are not reducible to the laws of physics and chemistry, and accepted the word "life" as a substitute for specific chemical information. The main casualty of his misconception was modern biochemistry, which was kept beyond reach for at least another thirty years.

The Birth of Modern Biochemistry

"Serendipity" is an overworked word, but there is no other that so well describes the discovery of cell-free fermentation or many other major advances in biology, including penicillin, DNA, and the genetic code. The word, coined by Horace Walpole in 1754, is based on a fairy tale, "The Three Princes of Serendip" (now Sri Lanka), the heroes of which "were always making discoveries by accidents and sagacity, of things they were not in quest of."

Eduard Buchner (1860–1917) worked as a teacher of chemistry at the University of Tübingen except during summers, when he spent his time in his brother's bacteriological laboratory in Munich. In 1897 he prepared an extract of yeast cells for a series of medical studies. As he described it:

> If one mixes 1000 grams of brewer's yeast with an equal weight of sand and then grinds the mixture, the mass becomes moist and pliable. Now if 100 grams of water are added, and the paste, wrapped in cheesecloth, is gradually subjected to 400–500 atmospheres of pressure in a hydraulic press, one obtains about 500 milliliters of 'press juice'; to remove any residual unbroken cells, the press juice is passed through a paper filter. The final solution contains a collection of substances derived from the cell interior. The 'cell extract' obtained in this way is a clear, slightly yellow liquid with a pleasant yeast smell.

To protect the juice from spoilage, he added sugar, as housewives did in preserving jams and jellies. To his dismay, bubbles of gas began to appear within a few minutes, and this went on for days. Instead of discarding the unpromising preparations, Buchner had the sagacity to wonder whether he might have accidentally witnessed something Pasteur had convinced most everyone was impossible, the cell-free fermentation of sugar to alcohol and carbon dioxide. Careful, repeated analyses convinced him this was so. "An apparatus as complicated as the living yeast cell" was not essential. "Rather a dissolved substance" was seen as "the carrier of the fermentation activity of the press juice." One could now search in the juice, freed from the constraints of condensed cellular architecture, for the substance, which he named *zymase* (Greek *zyme*, leaven), responsible for the chemistry of fermentation. Modern biochemistry was born.

Why had Buchner's simple experiment been so long delayed? The chemists were poised to do it for fifty years. In opposition to Pasteur, they had proposed correctly that the sugar molecules were destabilized by ferments (that is, enzymes) in grape juice and even conceded these ferments might have been released by the yeast. Moritz Traube (1826–1894), a German wine merchant with a private laboratory who had been trained in chemistry by Liebig, had stated clearly, more than twenty years before Buchner's serendipitous discovery, that enzymes in microbes are the agents that catalyze cellular chemistry, such as the yeast fermentation. He urged that the enzymes be isolated, with the admonition that failures might be anticipated due to alteration or inactivation of the unstable catalysts.

Pierre Berthelot (1827–1907) had already taken a major step in this direction. As early as 1860 he regarded action by invertase (an enzyme in yeast juice that converts sucrose to its component sugars glucose and fructose) as one of the many cellular operations that Pasteur argued were inseparable from the living cell. Pasteur himself, as a posthumous study of his records revealed, had prepared cell-free yeast juices but found them unable to carry out the fermentation process. His failure in these attempts confirms his oft-quoted aphorism: "In the field of experimentation, chance favors only the prepared mind." His mind, it seems, was unprepared for the success of a cell-free fermentation. He had also been unlucky in the strain of yeast he chose to use. Years later, it was learned that Berthelot's invertase, responsible for the first step in the fermentation of sucrose, does not survive the extraction procedure Pasteur applied to his Parisian yeast strain, whereas the enzyme in the Munich strain, as Buchner used it, apparently did survive.

Enzymes: The Vital Force

What chemical feature most clearly enables the living cell and organism to function, grow, and reproduce? Not the carbohydrate stored as starch in plants or glycogen in animals, nor the depots of fat. It is not the structural proteins that form muscle, elastic tissue, and the skeletal fabric. Nor is it DNA, the genetic material. Despite its glamor, DNA is simply the construction manual that directs the assembly of the cell's proteins. The DNA itself is lifeless, its language cold and austere. What gives the cell its life and personality are enzymes. They govern all body processes; malfunction of even one enzyme can be fatal. Nothing in nature is so tangible and vital to our lives as enzymes, and yet so poorly understood and appreciated by all but a few scientists.

To remove the confusion in using "ferment" to refer both to the whole yeast cell and to a catalytic substance extracted from it, Wilhelm Kühne in 1878 introduced the term *enzyme* (Greek, in yeast) for the catalyst. He meant the term also to include similar substances in plants and animals which, in his words, "are not so fundamentally different from the unicellular organisms as some people would have us believe." Even Kühne would have been surprised to learn how right he was about the ubiquity of enzymes and the unity of biochemistry in Nature.

Enzymes are protein molecules which make things happen to other molecules that would otherwise remain absolutely static. They make stable carbohydrates, fats, and proteins susceptible to digestion and then catalyze the use of the digestion products as building blocks to make new cells. Enzymes extract energy from food and sunlight and carry out the myriad of chemical operations that distinguish one cell from another, man from mouse, and health from disease. Why one cell becomes bone and another brain, one normal and another cancerous, are questions still without final solution because we know too little about our enzymes.

The first clues to the existence of enzymes came from curiosity about the digestion of food, fermentations of sugar, and leavening of bread. Two French scientists, Anselme Payen (1795–1871) and Jean Persoz (1805–1868), wondered about the germination of barley grain and found in 1833 that the germinated grain (or malt) exudes a substance which can break down a starch fiber into its many constituent glucose molecules. The concentrated substance, dried to a white powder and redissolved in water, could in ten minutes convert one thousand times its weight of starch into glucose. Thinking the substance caused the breakage of the envelope around the starch granules, Payen and Persoz called it *diastase*, after the Greek word for making a breach; the enzyme was renamed *amylase* when it became clear that its action was directly on amylose (Latin *amylum*, starch). Soon after, Schwann, looking for the "digestive principle" in stomach juice which degraded the proteins of fibrous tissues and muscle, found a material which he called *pepsin*. We also recall the discovery of *invertase* (sometimes called *sucrase*), the agent Berthelot extracted from yeast that splits the sucrose of grape juice into its component sugars glucose and fructose.

In naming each of the thousand and more enzymes that have been discovered, the suffix *ase* has usually been added to the substrate it affects or the chemical process it catalyzes. The name *zymase* that Buchner attached to the enzyme in yeast juice that converts sugar to alcohol and carbon dioxide was replaced when subsequent studies proved that this conversion required not one but a dozen distinctive enzymes. These enzymes act in sequence to direct intricate chemical operations that supply energy for the yeast cell's growth and multiplication.

Just what are enzymes and how do these little chemists do their remarkable work? Despite having worked with enzymes nearly all my life, I still find it hard to fathom their size. Just as the macroscopic dimensions of the universe exceed our mundane comprehension, so do the many powers of ten on the microscopic scale that we must apply to understand the size of molecules and their atoms (Fig. 2-2).

Enzymes, as proteins, are large molecules composed of a chain (or chains) of amino acids. Ten thousand atoms are needed to make several hundred of the twenty kinds of amino acids that make up a protein chain (Fig. 2-3). Each protein chain has a unique sequence of amino acids; the variety of possible chains, given this assortment of amino acids, is virtually unlimited. The particular sequence of amino acids in a protein molecule dictates how the chain will fold to give it the overall shape and surface features that are crucial for its function. A protein that functions as an enzyme may have a claw, cleft, hollow, or knob to grasp, hold, stretch, and bend the molecule it acts upon, which is called the *substrate*.

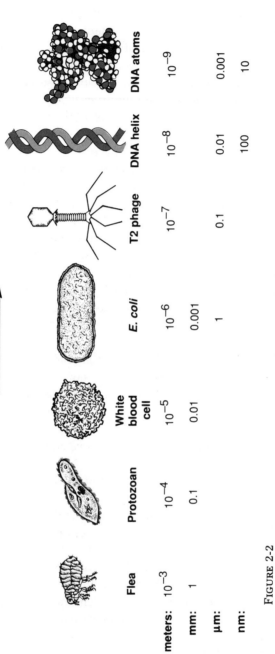

Six Powers of Ten: from a flea to DNA

	Flea	Protozoan	White blood cell	E. coli	T2 phage	DNA helix	DNA atoms
meters:	10^{-3}	10^{-4}	10^{-5}	10^{-6}	10^{-7}	10^{-8}	10^{-9}
mm:	1	0.1	0.01	0.001			
μm:				1	0.1	0.01	0.001
nm:						100	10

FIGURE 2-2
Atoms in a DNA molecule need to be magnified a million-fold to be as visible as a flea. (Sizes cited are approximate.)

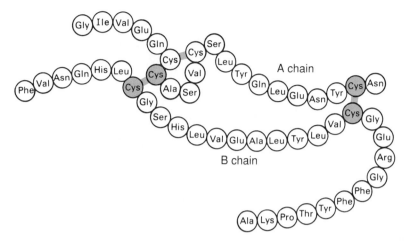

FIGURE 2-3
Insulin, a protein molecule, contains 18 of the 20 different kinds of amino acids in two distinctively ordered chains.

Sucrase is 400 times the size of its substrate, sucrose. It holds sucrose in such a way that the bond between atoms in sucrose's constituents (glucose and fructose) is strained and can be cleaved by the elements of the water molecules which surround it (Fig. 2-4). The union of enzyme with substrate and the chemical conversion that this union makes possible last a small fraction of a second. The products, in this case free glucose and fructose, then separate from the enzyme. Thus the general equation for an enzyme reaction is:

Substrate + Enzyme → *Substrate:Enzyme* → *Product:Enzyme*

→ Product + Enzyme

An enzyme has three remarkable features. It is

(1) *Catalytic.* It commonly increases the rate of a reaction 10-billion-fold. Thus, a reaction which would on average happen once in a thousand years is achieved by enzyme catalysis in one second. The enzyme molecule, as shown in the equation, is regenerated after each act of converting a substrate to a product. In other words, it speeds up the rate of the reaction, but it is not consumed or changed in the process.

(2) *Effective in tiny amounts.* One enzyme molecule commonly converts 1,000 molecules of substrate in a minute; some enzymes are known to convert 3,000,000 in a minute.

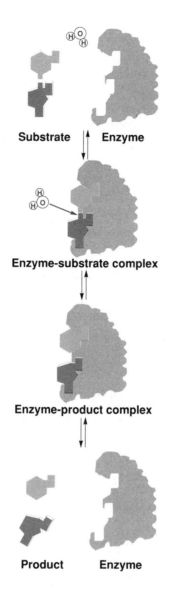

Substrate   **Enzyme**

Enzyme-substrate complex

Enzyme-product complex

Product   **Enzyme**

FIGURE 2-4
The enzyme sucrase cleaves sucrose into glucose and
fructose by causing a water molecule to be divided
between these two constituents.

(3) *Highly specific.* One kind of enzyme will selectively carry out only one of many possible chemical alterations to which a substrate may be susceptible.

These feats of catalytic potency and exquisite specificity, attained by enzymes during more than three billion years of evolution, are far beyond the laboratory skills of any chemist. Instead, the resourceful chemist relies on enzymes to perform analyses and syntheses which would otherwise be impossible. Virtually all genetic engineering depends on enzymes to create and rearrange DNA into novel genes and chromosomes.

Respiration in the Test Tube

Respiration to most people means the exchange of fresh and spent air in the lungs. In the nineteenth century it was discovered that the respiratory processes which consume oxygen and produce carbon dioxide take place inside cells throughout the body.

Not all cells require oxygen to live. Among the newly discovered microbes, some were observed to grow in the absence of air or oxygen. Pasteur showed that fermentation in the yeast cell, for example, is such an anaerobic (oxygen-free) process. Although life without oxygen is possible for yeast and many other microbes, it is far less fuel-efficient. A supply of oxygen makes life easier. A yeast cell in air consumes only a twentieth as much glucose for its growth as it does during anaerobic fermentation. That is because, using oxygen, the cell is able to produce and store about twenty times as much energy for each glucose molecule it consumes. In the presence of oxygen, complete combustion of one molecule of glucose yields 38 molecules of ATP (adenosine triphosphate, Fig. 2-5)—the organic compound which stores chemical energy for conversion to work, light, or electrical energy or the synthesis of cellular substance. This yield of ATP is in contrast with the anaerobic process, in which the incomplete combustion of glucose produces only 2 ATPs. The 70-percent efficiency of energy conversion in aerobic (oxygen-using) metabolism is more than twice that of the finest combustion engine devised by man.

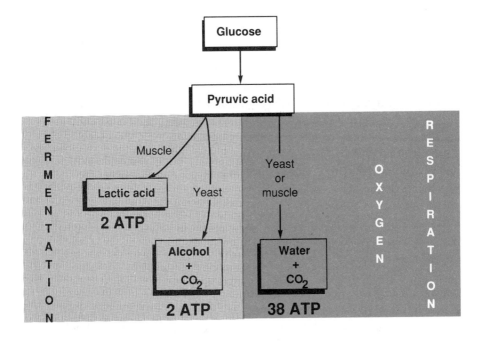

FIGURE 2-5
Metabolism of glucose in the absence of air (fermentation) captures only about a twentieth of the energy obtained by its combustion in air (respiration).

The outlines of the principal pathway of glucose metabolism were known in 1945. At the stage when a molecule of glucose (which has 6 carbon atoms) is rearranged to produce two pyruvic acid molecules (which have 3 carbon atoms each), the aerobic route diverges from alcohol production in yeast and from the anaerobic lactic acid production in muscle (Fig. 2-5). Instead, pyruvic acid is converted to a 2-carbon compound resembling acetic acid (vinegar), which is then fed through a complex series of reactions that convert it to carbon dioxide and water (Fig. 2-4).

A flow of hydrogen atoms and electrons in this citric acid cycle of combustion is linked to oxygen by red iron enzymes called *cytochromes*. Hydrogen and electron flow from one substance to another (for example, from pyruvic acid to oxygen) is called *oxidation;* the substance losing hydrogen or electrons is being oxidized. Simple examples of oxidation are the rusting of iron and the burning of natural gas (methane):

$$\text{Iron + oxygen} \rightarrow \text{Iron oxide}$$

$$\text{Methane + oxygen} \rightarrow \text{Carbon dioxide + water}$$

Iron and methane give up electrons to oxygen and are said to be "oxidized."

Just how ATP is generated in the course of pyruvic acid oxidation by oxygen was unknown in 1945. Where could I start to find out? I apprenticed myself to Bernard Horecker, a friend of mine at NIH. With the war over in the fall of 1945, he was about to return to the study of the cytochromes of cellular respiration. This had been the subject of his doctoral dissertation completed at the University of Chicago in 1940. At that time, unable to find any fellowship or research appointment, he had to take a desk job with the Civil Service Commission in Washington. Then a year later, an opening in the Industrial Hygiene Research Division of NIH had given him the chance to return to the laboratory. During the war years he had worked on how the insecticide DDT killed cockroaches. Now with the war over, he could resume his interest in enzymology and the cytochromes and was willing to have me work part-time in his laboratory.

We decided to examine one of the steps in the oxidation of pyruvic acid, a step in which a related compound, succinic acid, which has 4 carbon atoms, is oxidized with the mediation of cytochrome c. First, we had to isolate from finely minced calf heart muscle the proteins we needed to study this reaction: succinic acid oxidase, the enzyme that removes electrons from succinic acid, and cytochrome c, the small protein that transports the electrons on the way to oxygen to convert it—with the addition of two protons (H^+)—to water.

The details of how an electron is ferried in bucket-brigade fashion from succinic acid to oxygen (Fig. 2-6) is of great interest for several reasons. For one, the first bucket discloses the place where another of the B vitamins, riboflavin, carries out its vital function. Along the way, the electron is passed on to cytochrome c and thence to cytochrome oxidase, which finally, on giving it to oxygen, combines with hydrogen to form water. (It is at this last step that the poisonous action of cyanide operates by binding and blocking cytochrome

FIGURE 2-6

Electrons are passed in a "bucket brigade" from sugar-like molecules to oxygen, allowing energy to be captured at various steps along the way.

oxidase, thereby shutting off cellular respiration and killing the animal.)

By this elaborate electron bucket brigade the cell can, at several stages, capture the released energy in small, convenient packets as ATP. Without the succession of buckets, combustion of succinate would release chemical energy in a large burst of heat which could not be harnessed for cellular work or growth. If we think of ATP as the energy currency of a cell, the denomination of an ATP molecule is an amount that is readily negotiable. It is a convenient amount to pay for a large number of cellular functions; a molecule of larger denomination would be squandered most of the time. (Imagine having only thousand-dollar bills to do all of one's shopping.)

To monitor the progress of the enzyme's action on succinic acid in earlier decades, the starting material and the product would have to be isolated and weighed. Otto Warburg had replaced this cumbersome, laborious, and insensitive procedure with a convenient device which could determine small changes in gas pressure. The miniaturized manometer (gas pressure meter) which he introduced to measure the respiration of cells and tissues became the sine qua non of the enzymological laboratory in the 1930s and 1940s. Using this technique, we could measure the volume of oxygen consumed in the oxidation of succinic acid. I mastered the calibration and manipulation of the apparatus, but by 1945 it was already being eclipsed by a far more elegant device in which Warburg justifiably took greater pride.

The new device was a spectrophotometer. It was far superior to the manometer in speed, precision, and sensitivity. Light absorption at a particular wavelength (that is, color), including the invisible ultraviolet and infrared regions, could be measured continuously. Instead of determining the consumption of oxygen, we could observe the movement of an instrument needle as cytochrome c changed from blood-red (similar to the color of hemoglobin, the principal oxygen-carrying protein of blood) to violet-red as it carried electrons from succinate to oxygen. Horecker and I had access to the earliest Model DU spectrophotometer produced by Arnold Beckman's fledgling instrument company in California (Fig. 2-7). (Eventually, more than 30,000 of these instruments became fixtures in biochemical and chemical laboratories around the world.)

We routinely included cyanide in our assays to trap the electron bound to cytochrome c and, by its change in color, measure the rate

FIGURE 2-7
Severo Ochoa at a Beckman Model DU spectrophotometer, 1955.

at which it was reduced by succinic acid oxidase. (It became one of my jobs to measure out the cyanide solution. I could suck it into a pipet benignly unaware of its odor, which upset Horecker, whose genetic endowment, I presume, made him a thousand times more sensitive.)

Our probings, Horecker's and mine, with cytochrome c and its oxidation did not bring us to our goal, which was to discover the major source of ATP generation. But we were pleased to make an original observation along the way, which we thought worth describing in detail for publication in the *Journal of Biological Chemistry*. We found that cyanide, in addition to binding cytochrome oxidase, also attaches to cytochrome c, although the rate of that attachment is insignificant compared with its instantaneous action on cytochrome oxidase in blocking respiration.

We also included another datum in the publication: previously unreported in the scientific literature, cytochrome c absorbs light near the infrared region of the spectrum. This rather trivial item attracted the attention of Hugo Theorell, the most eminent Swedish enzymologist and later a Nobel laureate for his work on respiratory enzymes. Theorell had written a review for *Advances in Enzymology*, which Sylvy happened at that time to be editing for Academic Press, the publishers of this series. She came upon the statement that the Horecker-Kornberg claim of discovery of a new absorption band for cytochrome c was unwarranted because it had been described in 1934 by the Belgian biochemist E. J. Bigwood. When I told Horecker about this, he said he knew of the paper and that Bigwood's observation was not relevant. He wrote Theorell to that effect. We were mildly astonished to see his response, as it appeared in a "Letter to the Editors" of the *Archives of Biochemistry*:

In the *Advances in Enzymol.* 7, 281 (1947), I have stated that the ferricytochrome absorption band at 692 nm, described as a new observation by Horecker and Kornberg in 1946, was described by Bigwood *et al.* in 1934. Dr. Horecker has pointed out to me that Bigwood's bands at 675 and 640–45 mm appeared only in alkaline solution and thus cannot be the same as those appearing in neutral solution. Renewed study of the papers of Bigwood *et al.* has convinced me that Dr. Horecker is right in this respect. This, however, does not mean that Horecker and Kornberg are justified in claiming any priority on the discovery of the ferricytochrome bands at 695 and 655 nm in neutral solutions, since these bands were described by myself in 1940–43. I regret having credited Bigwood *et al.* with an observation made by myself. Horecker and Kornberg in their paper quote neither Bigwood's nor my papers.

Ten years later, when I got to know and like Theorell, I was tempted, but not bold enough, to twit him about the big fuss he once made over a scientific crumb.

Our enzyme preparations that oxidized succinic acid and consumed oxygen generated no ATP. This was no surprise. To capture the chemical energy of succinic acid combustion as ATP we needed a more complex system found in slices or granules of respiring tissues, such as muscle, liver, kidney, or brain. Enzymes free in solution did not work, and attempts to release from particulate matter the enzymes that coupled ATP synthesis to aerobic metabolism had failed. The enzymes of ATP synthesis would have to be separated from their organized cellular structure and purified to figure out how they worked. That seemed to me the most important thing I could do in biochemistry.

"Don't Waste Clean Thinking on Dirty Enzymes"

This admonition, which is often attributed to me but is original with Efraim Racker, is at the core of enzymology and good chemical practice. It says simply that detailed studies of how an enzyme catalyzes the conversion of one substance to another is generally a waste of time until the enzyme has been purified away from the other enzymes and substances that make up a crude cell extract. The mixture of thousands of different enzymes released from a disrupted liver, yeast, or bacterial cell likely contains several that direct other rearrangements of the starting material and the product of the particular enzyme's action. Only when we have purified the enzyme to the point that no other enzymes can be detected can we feel assured that a single type of enzyme molecule directs the conversion of substance A to substance B, and does nothing more. Only then can we learn how the enzyme does its work.

The rewards for the labor of purifying an enzyme were laid out in a series of inspirational papers by Otto Warburg in the 1930s. From his laboratory in Berlin-Dahlem came the pioneering methods and discipline of purifying enzymes and with those the clarification of key reactions and vitamin functions in the fermentation of glucose. Warburg was the *Geheimrat*, the dominant and domineering chief of his laboratory. (Hitler, through his fear of cancer, was persuaded to make Warburg an honorary Aryan. Despite his Jewish ancestry, Warburg could boast, after the war and the Holocaust, "I have been working uninterruptedly in Berlin-Dahlem, disturbed by nobody, not even by dictators.") One of his well-trained assistants, Erwin Haas, came as a refugee to Chicago, where he introduced Bernie Horecker, who in turn introduced me, to the ways of enzyme purification.

My brief experience in Horecker's lab whetted my appetite to learn more enzymology. Having prevailed on Henry Sebrell, the chief of my division, to get permission for me to go to another laboratory for research training (an unprecedented fellowship arrangement for a commissioned medical officer at NIH), where was I to go in 1945? Germany, the Mecca of the 1930s, was now an anathema and ruined. (Forty years later—a whole generation of scientists having been destroyed—biochemistry in Germany has not yet recovered.) I was attracted to Cambridge, England, with its many famous names, especially David Keilin, whose work on the cytochromes was preeminent. When I consulted David Green, who had returned to the United States after many years of productive work on enzymes in Cambridge, he advised me against going there. As a result of the war, the laboratories were in poor shape and the scientists dispersed. Instead, he urged that I join his laboratory at Columbia University. I was somehow put off by his manner and was then made aware by friends of another alternative, a young Spanish biochemist working at New York University Medical School.

Two years earlier, Severo Ochoa (Fig. 2-7) had demonstrated for the first time that pigeon brain particles make 3 ATPs for every atom of oxygen consumed in the combustion of pyruvic acid. He was now purifying enzymes in this metabolic pathway that might reveal the mechanism of this coupling of ATP synthesis to oxidation, which he had called *oxidative phosphorylation*. He was doing the very kind of thing I hoped to do and was described by friends as a kind and enthusiastic person. A scouting report on his background was reassuring.

Ochoa, born in Spain in 1905, had obtained a medical degree, generally the only course available then in Europe to pursue an interest in biology. Except for brief stays in Madrid between 1931 and 1935, he had worked abroad in leading biochemistry laboratories: in Berlin, and later in Heidelberg, with Otto Meyerhof; in Oxford with Rudolph Peters; and in St. Louis with Carl Cori. Buffeted about by the turbulence and wars in Spain, Germany, and England, he was now struggling to find a haven in the New World. Through-out, he remained steadfast and confident in his devotion to science.

He was working alone in laboratory space assigned to Professor Isidor Greenwald in NYU's Department of Biochemistry after having been evicted from provisional space in the Psychiatry Department in Bellevue Hospital across the street. (On returning to the laboratory from a concert one Sunday afternoon, he had found his desk and equipment moved out into the hall.) Greenwald told Joseph Bunim, an internist on the faculty and a friend of my family, that Ochoa impressed him as a serious and able scientist. I went to New York to meet Ochoa in November and found a courtly, charming, El Greco figure excited about his work. He agreed to take me, and I started, as his first postdoctoral student, a month later.

That year in New York was one of the happiest and most exhilarating in my life. Never had I learned so much so quickly. And in the few waking hours outside the laboratory, Sylvy and I discovered the theater, music, and museums that are the heart-throb of New York. Despite my being a native of Brooklyn and having attended City College in uptown Manhattan and despite Sylvy's many visits from Rochester, where she grew up and studied biochemistry, we were strangers to the city. We were also homeless. Vacant apartments in New York right after the war were practically nonexistent.

It was Herman Kalckar who saved us, the same Kalckar to whom we owe the earliest insights into ATP and the energetics of respiration. He had met Sylvy in Rochester and liked her. Learning of our plight at a time when he was planning to return to Denmark, with his unfailing thoughtfulness and generosity he intervened with his landlord to let us take over his apartment on West 123rd Street, overlooking Morningside Park. (The apartment house has since been replaced by a housing project.) My naval uniform may also have helped persuade the landlord. There was one hitch, however. The Kalckars would not be leaving until mid-February. For six weeks Sylvy and I moved, often daily, from one hotel to another, including the best and the sleaziest. We couldn't afford a fine hotel for more than a day; a modest one would make us move after five days to avoid giving us the rent-controlled weekly rate. A common routine for Sylvy was to check out of a hotel in the morning and wait in line, sometimes all day, at the Officers Service Club for a new hotel assignment. She would then call me at the lab to tell me where we would sleep that night. Some nights we were awakened by roaches or vermin or nearby bedlam. One morning we overslept badly because the room had no window.

On December 26, 1945, I arrived in Ochoa's lab, completely committed to learning biochemistry and finding the mother lode of ATP. Waiting for me were six fresh pig hearts. Highly contractile muscles, such as those of the heart or a pigeon's breast, are attractive places to find an abundance of enzymes responsible for respiration and ATP synthesis. Ochoa suggested I use these hearts as a source to purify aconitase, one of the enzymes operating in the pathway of pyruvate oxidation. He embarked on the purification of enzymes in this pathway of sequential reactions because of his conviction that an understanding of the mechanism of oxidative phosphorylation required knowledge of each of the enzymatic events coupled to phosphorylation.

My mission was to purify heart muscle aconitase. We expected to resolve the activity into the two enzymes that account for the successive subtraction and readdition of a water molecule (in a different position) that converts citric to isocitric acid. This was my first solo stab at enzyme purification, and after a few months' work I failed to make any significant progress. I also found no signs of separation of the expected two enzymes responsible for the two reactions; aconitase, as others showed much later, is indeed a single enzyme.

Despite initial failures, this immersion in enzymology was intoxicating. Aside from the fascination of seeing an enzyme in action, the momentum of experimental work was breathtaking. By coupling aconitase action to the next enzymatic step, assays could be carried out in the spectrophotometer. Aconitase conversion of citric to isocitric acid leads to no spectral change, but removal of isocitric acid as soon as it is formed (by another enzyme that we added to the reaction mixture) can be seen and measured precisely. Because the assays took only a few minutes, many ideas could be tested and discarded in the course of a day. Late evenings were occupied preparing a series of protocols for the following day. What a contrast with the tedious pace of nutritional experiments on rats!

In my work on aconitase, I learned the philosophy and practice of enzyme purification. To attain the goal of a pure protein, the cardinal rule is that the ratio of enzyme activity to the total protein is increased to the limit. Units of activity and amounts of protein must be strictly accounted for in each manipulation and at every stage. In this vein, the notebook record of an enzyme purification should withstand the scrutiny of an auditor or bank examiner. Not that I ever regarded the enterprise as a business or banking operation. Rather, it often seemed like the ascent of an uncharted mountain: the logistics resembled supplying successively higher base camps; protein fatalities and confusing contaminants resembled the adventure of unexpected storms and hardships. Gratifying views along the way fed the anticipation of what would be seen from the top. The ultimate reward of a pure enzyme was tantamount to the unobstructed and commanding view from the summit. Beyond the grand vista and thrill of being there first, there was no need for descent, but rather the prospect of even more inviting mountains, each with the promise of even grander views.

Pig Hearts and Pigeon Livers

We used fresh heart muscle obtained from pigs at the slaughter house or the breast muscle of pigeons, supplied by local dealers. The vigorous metabolism of pigeons made their livers also a good source of respiratory enzymes. Years later, we came to appreciate that bacteria, whose genetics, nutrition, and life cycle we could rigorously control—and so much more easily and more cheaply too—were superior sources for most biochemical operations.

My experimental notes of December 26 and 27 convey some sense of how purification of aconitase from the pig hearts was started (Fig. 2-8) and how the activity of the enzyme was measured (assayed):

> Fat and connective tissue removed. Muscle put through a meat grinder four times. Suspended the mince (968 grams) in 1.5 liters of ice-cold water to leach out activity. [Worked in the 'cold room' (ice temperature) and did everything in the cold.] Pressed the suspension through gauze and clarified the fluid in centrifuge. To the clear extract, added 540 grams of ammonium sulfate salt (55 percent of the maximum that could be dissolved in this volume of ice-cold water). The solid material that developed was separated in a centrifuge. Contained little aconitase activity (see assays) and discarded. To supernatant fluid, added 92 grams ammonium sulfate. Solid material collected in centrifuge and dissolved in water.

This particular fraction of proteins, made insoluble by ammonium sulfate, contained most of the aconitase activity in the original mus-

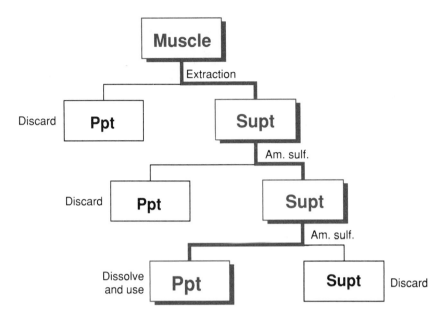

FIGURE 2-8
First steps in purifying the enzyme aconitase from a liquid extract of the
heart muscle of a pig. *Ppt*, precipitate of proteins rendered insoluble by
the addition of ammonium sulfate; *supt*, supernatant, or liquid phase, still
containing some proteins in solution.

cle extract but far less of the proteins. This standard technique of
fractionating proteins had exploited the differences in their solubil-
ity in ammonium sulfate solutions and enabled me to enrich (purify)
aconitase relative to the crude extract.

Our next problem was to figure out how to measure aconitase
activity. Warburg had discovered in another context that, in some
oxidations, a small molecule served as a *coenzyme*—a detachable
working part of an enzyme. He was astonished to find that the coenzyme contained a well-known chemical, nicotinic acid (or niacin),
an oxidation product of nicotine found in tobacco and other natural
substances. The coenzyme, called NADP for "nicotinamide adenine
dinucleotide phosphate," is considerably more complex than niacin.
It contains most of the ATP molecule, but niacin (N) is the active
ingredient (Fig. 2-9).

The niacin coenzyme provides an essential link in the electron
bucket brigade shuttling electrons from the fuel substance (for ex-

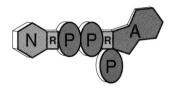

FIGURE 2-9
The coenzyme NADP is made up of Nicotinamide–Ribose–Phosphate
linked to Adenine–Ribose–Phosphate, but with an extra Phosphate attached.

ample, isocitric acid) to the cytochromes and thence to oxygen. Warburg made the extraordinarily useful discovery that the niacin coenzyme, upon receiving electrons, becomes visible in a particular
region of the ultraviolet spectrum. This color change (absorption of
light) of NADP upon reduction, like that of cytochrome *c* which is
visible to the eye, provides a precise measure of the number of
molecules or amount of the compound (in this case isocitric acid)
which has furnished the electrons.

The revelation by Warburg that niacin, in a more complex form, is
the substance responsible for a key event in cellular respiration led
others quickly to the discovery that niacin is the vitamin needed to
prevent or cure blacktongue in dogs and pellagra in humans. Yet,
today, even though we know exactly how niacin operates in all cells,
it is still a mystery why humans deficient in this vitamin suffer
chiefly from skin lesions, an intestinal disorder, and dementia. In the
same vein, it remains unclear why the lack of Vitamin B1 (thiamine),
a coenzyme in the first step of pyruvic acid metabolism in all tissues,
shows itself most prominently as the neuritis of beriberi.

Back to the assay of aconitase. The aconitase conversion of citric
acid to isocitric acid (in the citric acid metabolic cycle) was measured, as mentioned earlier, by coupling it to a subsequent reaction.
The moment isocitric acid was produced, it was instantly oxidized
by a generous quantity of another purified enzyme (isocitric dehydrogenase), which we routinely included in the assay mixture. Except for aconitase, the reaction mixture contained all the ingredients
in excess so as not to limit the reaction. The color changes observed
would then depend only on the amount of aconitase present. The
volume was 3 milliliters, a tenth of a fluid ounce; nowadays our
standard assays have been miniaturized to less than one-hundredth
this volume, a barely visible droplet.

I recorded the movement of a dial of the spectrophotometer every fifteen seconds, starting immediately after adding the solution being assayed for aconitase. There was a steady increase in UV light absorption (Fig. 2-10). The rates were proportional to the amounts of enzyme added. With reaction mixtures lacking one of the ingredients (aconitase, citric acid, NADP, or isocitric dehydrogenase) serving as controls, there was no change whatever, thus proving that the spectrophotometric change was indeed the result of the reaction we intended to observe.

To express the activity of an enzyme, we define an arbitrary unit, in this case, an increase in light absorption of 0.01 optical density unit per minute. On the basis of the assay (Fig. 2-10), I concluded that my solution contained 20 aconitase units in 0.02 ml or 1,000 units per ml. Protein content was determined by a color test using a protein stain in comparison with a known concentration of blood albumin from cows as a standard of reference. The results of these determinations provided the first two lines of the "bank account,"

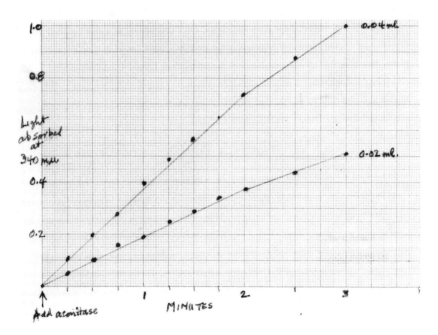

FIGURE 2-10
A spectrophotometric assay of aconitase taken from my notebook (February 1946).

TABLE 2-1
Purification of aconitase

Fraction	Enzyme activity (units)	Protein content (mg)	Yield of activity (%)	Extent of purification (units/mg protein)
Muscle extract	53,000	8,100	(100)	6.5
Ammonium sulfate	40,000	1,920	75	21

the "Purification Table," concisely stating the route taken toward isolation of a pure protein (see Table 2-1).

The values in the table for the aconitase activity and protein content of the muscle extract and the fraction derived from it indicate that we had recovered 75 percent of the activity and only 24 percent of the protein, and so had enriched our activity relative to protein about 3-fold. This modest progress was far short of what I had hoped for. A proper purification of aconitase came only years later, when many new and powerful methods had been devised to fractionate proteins based on their size, shape, electric charge, and chemical affinities. Purification of the muscle extract by about 100-fold turned out to be what was eventually needed to obtain a homogeneous (pure) preparation of aconitase, which, interestingly, contained an iron atom essential for its action.

In these six months of trying to learn about enzymes, I felt keenly my inadequacy in organic and physical chemistry and decided to take summer courses offered at Columbia University. In this first postwar summer, the condensed year-length courses were taught by the best faculty. Organic chemistry had been transformed in the twelve-year interval since I took the course at City College. In Linus Pauling's *Nature of the Chemical Bond* and other texts, the sticks that linked the ball-like atoms had been replaced by meaningful pairs of electrons. Pushing these electrons here or there rearranged the atoms to make smaller, larger, and novel molecules. An enzyme could be understood simply as a superior chemist selecting one or another of several possible reactions and directing it with extraordinary facility and precision.

Physical chemistry—the laws of kinetics and thermodynamics applied to molecules—makes mathematical statements of chemical events. To grasp these treasured levels of biological expression, I needed for the first time to understand energy and entropy. Energy, in its several interconvertible forms, seemed quite clear. But entropy, the disorder or randomness of a system, was intuitively less so. I came to realize that being familiar with a thermometer made the difference. The lack of a corresponding instrument to measure entropy makes it more difficult to develop a comparable intuitive sense of this property.

I found it strange in taking these optional courses that the dread of exams returned, along with the familiar nightmares of entering the examination room totally unprepared. And the grades, which I would share only with my wife, provoked the same chills and apprehensions I knew in my intensely competitive school and college years. Despite taking the courses, I never gained the fluent and unaccented command of these basic chemical disciplines that I might have enjoyed had they been part of an early curriculum.

Returning to the lab, I was luckier in my second attempt at enzyme purification, joining Ochoa and Alan Mehler, his first graduate student, in studies of a liver enzyme that acts upon malic acid, related to pyruvic acid. (This "malic enzyme" is also crucial in the malolactic fermentation of fine Bordeaux and Burgundy wines and California Cabernet Sauvignon.) Mehler was already on the scene when I arrived in Ochoa's lab and became my indefatigable and devoted tutor. Having always been the youngest in my class, I was shocked to find that I was so far behind someone four years my junior.

Mehler was inclined to put off doing an experiment until he could account for every feature and anticipate every possible outcome. His speculations taught me a lot, but they often led him to ponder for hours about an experiment whose result could be known in minutes. One such occasion was the only time I ever saw Ochoa's imperturbable good humor and serenity shaken. To one of Mehler's invariable Talmudic questions about a routine practice: "Is this really necessary?" Ochoa, red-faced and arms waving, sputtered: "It's necessary, it's necessary, yes, it's really necessary!"

At the end of 1946, just before I left Ochoa's laboratory and started for St. Louis, we undertook a very large-scale preparation of the malic enzyme, starting with several hundred pigeons who had donated their livers to Morton Schneider, Ochoa's talented and devoted assistant. Four of us had worked for several weeks to reach the last step in which successive additions of ethanol finally yielded a protein precipitate which we believed, from small-scale trials, would have the enzyme in an adequate state of purity. We had only to fill in these details in a paper we had prepared for publication. Rather late one evening, Ochoa and I were dissolving the desired enzyme which had been collected in many glass centrifuge bottles. I had just poured the dissolved contents of the last bottle into a measuring cylinder when I brushed against and overturned one of the wobbly bottles on the crowded bench. That bottle knocked over another and the domino effect reached the cylinder; it fell, and all of the precious solution spilled on the floor. Gone forever.

By the time I got home an hour later, Ochoa had already called. I had been so terribly upset that he was concerned about my safety. Back in the lab early the next morning, I glanced at the supernatant fluid separated by centrifuging the last fraction. I might have dumped it but had instead stored it at $-15°C$. The fluid had become slightly turbid, and I decided to collect the solid material in the centrifuge, dissolve, and assay it. Holy Toledo! This fraction had the bulk of the enzyme activity and was several-fold purer than the best of our previous preparations. This step (without the cylinder breakage) became part of the published procedure.

Chapter 3

Never a Dull Enzyme

Although the yeast cell can live and grow in the absence of oxygen, the use of fuel is incomplete. Alcohol is not oxidized to carbon dioxide and water, and so it accumulates. Life in air (oxygen) is far richer, and the number of ATP molecules generated from combustion of a glucose molecule is catapulted from two to thirty-eight. How this bonanza of ATP is created, the central question in biochemistry of the 1940s, was what first attracted me to the study of enzymes.

After fifty years of studies by hundreds of biochemists, fermentation of sugar to alcohol proved to be a series of twelve reactions of unanticipated complexity carried out by at least that many enzymes. To the student now saddled with this burden of knowledge, one can only sigh and quip: "I have yet to see any problem, however complicated, which when looked at in the right way, does not become still more complicated." A recent example from studies of one of the apparently simple steps in the pathway from sucrose to alcohol, a step catalyzed by one enzyme, aldolase, shows it to be further divisible into nine discrete chemical stages. Yet, the welter of biological phenomena I struggled to learn as a student has been given a gratifying coherence and clarity by its expression in chemical language.

Just as intriguing to the early biochemists as the fermentation of sugar to alcohol by yeast was the conversion of sugar to lactic acid by microbes making cheese and by muscles doing vigorous work. What is fascinating about each of these processes, beyond revealing chemical rearrangements of great novelty, is how the conversion of sugar provides energy for the yeast cell or other microbes to grow and for muscles to contract.

ATP, the Energy Currency of the Cell

Since the late eighteenth century it had been known that sugar, burned as a fuel, behaves like a candle or coal. The carbon atoms of sugar, paraffin, or coal unite with oxygen to form carbon dioxide.

$$\text{Carbon} + \text{Oxygen} \rightarrow \text{Carbon dioxide} + \text{heat}$$
$$\text{(C)} \qquad \text{(O}_2) \qquad\qquad \text{(CO}_2)$$

By this combustion, the chemical energy stored in the carbon becomes heat and light energy. The novel understanding reached around 1940 was that the cellular combustion of sugar, instead of wasting the energy as heat and light, captures much of the chemical energy of the sugar and stores it in another form—ATP—for later use. How does the ATP molecule store energy? How is the energy used in the manufacture of even more complex chemicals, such as proteins and DNA? How is chemical energy converted to mechanical energy for movement? Beyond these questions are other complexities: How does the yeast cell grow and multiply on energy derived from sugar even in the absence of oxygen, and how does a muscle keep contracting even after the oxygen is depleted? How can combustion and life processes be managed without oxygen?

Eduard Buchner's report in 1899 that yeast juice could sustain the fermentation of sugar to alcohol opened the way for others to explore the details of this reaction. Among them, Arthur Harden of London made the most significant discovery. Observing that only 5 to 10 percent of the added sugar was consumed by yeast-juice preparations, he naturally wondered why the process stopped so far short of completion. Perhaps the fermentation enzyme in the juice was unstable and decayed rapidly. Harden wondered whether aged and boiled yeast juice might prevent this destruction. For what proved to be the wrong reasons, he did the right experiment of examining the properties of boiled yeast juice and finding that it greatly enhanced sugar utilization.

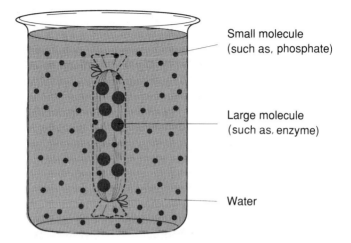

FIGURE 3-1
Dialysis in a cellophane sac retains enzyme molecules
while permitting the free passage of small molecules.

The experiment worked because the boiled juice (lacking active enzymes) provided additional amounts of certain small molecules which, it was later learned, are essential to sustain the action of the enzymes of fermentation. Harden discovered these small molecules by the simple device of dialysis, a filtration procedure (Fig. 3-1). Separation of these small molecules from enzymes can be effected by enclosing the yeast juice in a cellophane tube, similar to a sausage casing. Tiny pores in the casing allowed small molecules to pass out into a surrounding solution, but the openings were small enough to prevent the passage and loss of large molecules, such as enzymes.

One of the essential small molecules that passed through the dialysis membrane was identified by Harden as phosphate, the salt form of phosphoric acid. The others were not identified for many years. Why is phosphate so important in the fermentation of sugar? It turns out that phosphate is the crucial mineral of metabolism in every cell in Nature. Phosphate is the basis, directly or indirectly, for all energy negotiations, the "business end" of the fabulous ATP molecule. (Phosphate also forms the backbone of genetic material, the fabric of bone, and is now being seen as the way of regulating the proteins responsible for most biological processes, including brain function, growth, and cancer.)

Phosphate is incorporated into ATP after being attached to a part of the glucose molecule at a certain step during its combustion. A molecule of ATP contains a chain of three phosphate groups, which are linked to an adenosine molecule. Adenosine in turn is composed of two parts—ribose, a sugar, and adenine, also one of the building blocks of the nucleic acids (Fig. 3-2).

Chemical energy stored in ATP can be used to build complex molecules, contract muscles, generate electricity in nerves, and light up fireflies. All fuel sources of Nature, all foodstuffs of living things, produce ATP, which in turn powers virtually every activity of the cell and organism. Imagine the metabolic confusion if this were not so: each of the diverse foodstuffs would generate different energy currencies and each of the great variety of cellular functions would have to trade in its unique currency (Fig. 3-3).

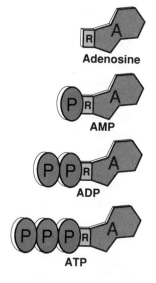

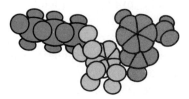

FIGURE 3-2
Adenosine (adenine + ribose), AMP (adenosine monophosphate), ADP (adenosine diphosphate), and ATP (adenosine triphosphate). A space-filling model of ATP (bottom) shows the atoms of carbon, hydrogen, nitrogen, oxygen, and phosphorus.

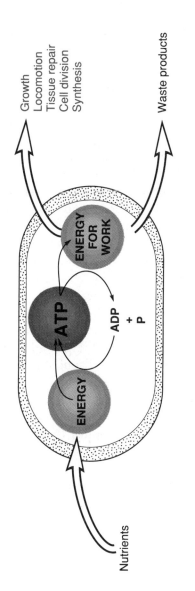

FIGURE 3-3
Energy for synthesis of ATP (from ADP and phosphate) comes from the combustion of food. Energy stored in ATP is used for mechanical, chemical, and electrical cellular functions; some energy is lost as heat.

In a typical energy transaction, ATP is made when phosphate combines with ADP (adenosine diphosphate—adenosine with two phosphates instead of three). The energy for synthesis of ATP is derived from the combustion or rearrangement of a molecule, such as sugar or fat. Then, when the stored ATP is cashed for an energy need of the cell, one phosphate is removed and ADP is produced again, releasing energy in the process. In this cycle of deposits and withdrawals of ATP, some discounting is inevitable. As with all mechanical and chemical processes, some energy is dissipated as heat, but the efficiency of the biological conversion of fuel to cellular work is astonishingly high.

How does the conversion of glucose to alcohol in yeast or to lactic acid in muscle generate ATP? The chemical details of the many reactions became known as one after another of the enzymes responsible for each was isolated in pure form and examined separately from the rest. In certain of these reactions, some of the B vitamins (niacin, the antipellagra vitamin, and thiamine, the beriberi-preventive vitamin) perform their essential functions.

Simply stated, a molecule of glucose is activated with an investment of two ATP molecules. Successive transfers of phosphate from ATP form a sugar diphosphate. The latter undergoes several rearrangements on the way to alcohol or lactic acid, in the course of which *four* ATP molecules are generated (Fig. 2-5).

Two ATP molecules are invested and four ATP molecules are produced. The net gain of two ATPs per glucose molecule consumed is sufficient for the yeast cell to grow and for a muscle to work. On the pathway from glucose either to alcohol and carbon dioxide or to lactic acid, the conservation of matter applies: there is no loss or gain of carbon, hydrogen, or oxygen.

In addition to the universality of ATP as the central molecule of energy transactions, the intricate mechanisms of ATP generation from glucose in both yeast and muscle proved to be virtually identical! The pathways in yeast and muscle diverge only at the end, when pyruvic acid produces alcohol and carbon dioxide in yeast but lactic acid in muscle. The enzymes, vitamins, and minerals are nearly the same in structure and function. The unity of biochemistry in Nature, first seen clearly in these studies, applies not only to glucose metabolism but also to the other basic features of cellular operations. From the genetic code to the encyclopedic details of the thousands of cellular proteins, these evolutionary triumphs achieved 2 to 3 billion years ago have withstood countless encounters with forces that could mutate or replace them. At these fundamental levels, there are few distinctions between a human, a mouse, and a yeast cell.

A staggering amount of ATP is turned over each day to convert the energy in food to the forms we need to live and work. The average daily intake of 2,500 calories translates to the turnover of about 180 kilograms (400 pounds) of ATP! With the body content of ATP being only 50 grams (about one-tenth of a pound), the cycling of ATP—its synthesis from ADP and phosphate and its subsequent breakdown—must occur about 4,000 times a day. Expressed another way, 50 grams of ATP contains 60 billion trillion molecules (6×10^{22}), in each of which, on the average, the terminal phosphate is added and removed 3 times each minute. In 1945, the mechanisms responsible for 95 percent of this ATP synthesis were totally unknown.

The inefficient use of sugar in the absence of oxygen, first noted by Pasteur in 1876, has intrigued many biochemists, including Otto Warburg, whom some regard as the greatest of them all. In the 1920s before ATP was discovered, he observed something unusual: Cancer cells and embryonic tissues accumulated lactic acid even in the presence of oxygen. He sought, as did other investigators, an enzyme that might normally switch the gears from anaerobic to aerobic metabolism when oxygen becomes available. Such a simple enzyme, whose inactivation might be basic to the cancer process, has never been found. The switching mechanism is enmeshed in an intricate network of reactions that enables the energy of the complete combustion of glucose to be largely conserved as ATP. The nature of this startlingly efficient device and how it operates in normal cells to generate an exceptionally rich deposit of ATP was what I was after when I entered biochemistry in 1945.

On the Trail of the ATP Bonanza

The mother lode of ATP production had been discovered in 1939 in studies of thin slices or particles of tissues, such as kidney, liver, brain, or muscle. Herman Kalckar, then in Denmark, was among the first to become aware of this extraordinarily efficient synthesis of ATP, called aerobic or oxidative phosphorylation. In his first visit to the United States in 1939, on his way to the California Institute of

Technology in Pasadena, Kalckar stopped in St. Louis to visit Carl and Gerty Cori at Washington University Medical School. Having been unable to repeat his experiments, published two years earlier, Gerty was all for asking Kalckar for a demonstration, but Carl thought it an improper imposition on a visitor. They settled instead on having Sidney Colowick, their graduate student, describe in detail how he had gone about doing the Kalckar experiment. (Of several celebrated papers authored by Cori, Colowick, and Cori, the wry Colowick once remarked: "I'm the meat in the Cori sandwich.")

Colowick explained how he assembled the ingredients, just as Kalckar described them. He of course used test tubes, as he and the Coris were accustomed to doing in their classic studies of muscle enzymes. Kalckar, listening to Colowick's account of failed attempts to observe aerobic phosphorylation, immediately saw the problem. "You must keep shaking the mixture, Sidney," he said. "The kidney tissue has a craving for oxygen." Constant aeration in a shallow vessel was the answer.

How was I to find the fountain of ATP in a thin slice of kidney? Many years earlier Otto Warburg had concluded: "Where structure begins, biochemistry ends." By this dictum, he meant that purification of enzymes, essential for progress in biochemistry, cannot proceed when the enzymes remain locked in particulate, indissoluble structures. I had not understood this, and so I wondered why he and the other major figures in enzymology in the 1940s had turned away from the search for the enzymes that captured the energy as ATP in the aerobic combustion of body fuels. Determined to work on this problem, I had persuaded Henry Sebrell, and he in turn the Director of NIH, to let me extend my stay with Ochoa to a full year and then to spend six months more in Carl and Gerty's laboratory. Right after the war, it was the liveliest enzymology lab in the world.

St. Louis was the farthest west either Sylvy or I had ever been. We arrived in January of 1947, and this time, through the efforts of our Rochester friends Ralph and Esther Woolf, we had a place to stay. We were the first occupants of a room built in the cellar of an old house in a neighborhood near the laboratory. It was a dim and warm burrow next to the house furnace, a cozy nest for our first child, Roger, who was born in April.

The Cori laboratory was in the center of the biochemical spotlight in the 1940s. That this spotlight moves so quickly, darkening gigantic figures like Hopkins and Warburg, disappoints me. My students now are totally unaware of the work and spirit of Carl and Gerty Cori and of their guiding influence on my generation of biochemists.

Carl Cori was born in 1896 in Prague, then part of Austria, to a family with academic and scientific stature. His interests in biology led him to medical school, where he met and married a classmate, Gerty Radnitz. The war and its aftermath in Central Europe drove them in 1922 to America to pursue science. In Buffalo, at the State Institute for the Study of Malignant Disease, now known as the Roswell Park Memorial Institute, they began their work on the metabolism of carbohydrates. In 1931 this work attracted an invitation from Washington University for Carl to become Professor of Pharmacology in the School of Medicine. Professorial appointments at that time were scarce and prestigious, and the medical school at Washington University had few peers in its support of basic research in medicine. Gerty had no faculty appointment until 1947, when she shared the Nobel Prize with Carl (Fig. 3-4).

Carl and Gerty had the experimental gifts and wisdom to extend physiologic studies of whole animals and organs to crude cell extracts and, finally, to purified enzymes. They isolated glycogen phosphorylase in the form of a pure, crystalline protein. This enzyme mobilizes glycogen, the polymerized form of glucose stored in liver and muscle, to produce an activated form of glucose for metabolism. While teaching themselves and their students how to examine the molecular shapes and operations of enzymes, they never relinquished their devotion to understanding the function and hormonal control of enzymes in the intact animal.

At Washington University I was assigned to John Taylor, an assistant professor, perhaps as Cori's reward to him for his service as administrator of departmental business. Taylor was experienced in protein chemistry and was prepared to guide me in attempts to crystallize lactic (lactate) dehydrogenase, the enzyme that interconverts pyruvic and lactic acids; production of lactic acid is a crucial step in the metabolism of glucose in a vigorously contracting muscle. Key enzymes in the conversion of glycogen to pyruvic acid had already been obtained in crystalline form from rabbit muscle by the Coris in previous years. As a by-product of those preparations, Gerty Cori had accumulated a paste of proteins precipitated by ammonium sulfate. It was an excellent starting material for me to obtain in two weeks a near homogeneous preparation of lactic dehydrogenase.

I had no strong interest in crystallizing lactic dehydrogenase and

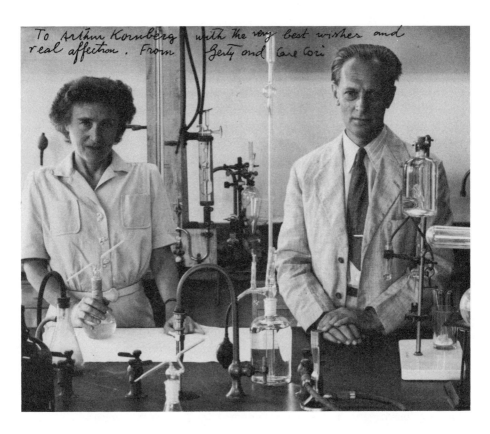

FIGURE 3-4
Gerty T. Cori and Carl F. Cori (circa 1947).

rebelled at continuing. I had come to the Cori laboratory to solve the major problem in biochemistry, the mechanism of aerobic phosphorylation. I had already lost nearly a month. It did not discourage me that Cori, Ochoa, Kalckar, and Lipmann, each of whom had contributed so much to the recognition of aerobic phosphorylation, had given up working on the problem. They were working on soluble enzymes rather than struggling with the particulate suspensions which sustain aerobic phosphorylation; they were adhering to the Warburg dictum.

Cori was remarkably tolerant and suggested I join a young Swedish visitor, Olov Lindberg, who was pursuing a striking observation made about six years earlier by Ochoa when he worked with the Coris. Liver particles metabolizing pyruvic and related acids produced inorganic (lacking carbon) pyrophosphate, a compound previously unknown as a cellular constituent. Inorganic pyrophosphate had been obtained by fusing two phosphate molecules through dehydration in an oven at 400°C! (The chemical energy trapped in pyrophosphate is comparable to that trapped in ATP from the fusion of ADP with phosphate.)

The mechanism of the origin of inorganic pyrophosphate in the liver preparations was unknown. One possibility excited me. Perhaps it was released from a very unstable form of ATP during isolation of the pyrophosphate from the reaction mixture. This novel "super ATP" might be a crucial intermediate in the mechanism of aerobic phosphorylation. How funny that I should now be working on the origin of pyrophosphate. During the previous year with Ochoa, he had mentioned the pyrophosphate phenomenon several times at our daily lunch with Racker and Mehler in the dining room we called Salmonella Hall. The strangeness of pyrophosphate and the mystery of its origin made it difficult for us to retain the details of the experiments, and we would pester him to repeat the story. Even Ochoa's patience could be exhausted, and finally on one occasion he forbade any further mention of the inorganic pyrophosphate business.

Lindberg and I ground up the metabolically active outer layers of rabbit kidneys and measured how well these preparations used various compounds related to pyruvic acid. As oxygen was consumed, phosphate was assimilated into ATP and, in confirmation of Ochoa's observation, pyrophosphate also accumulated, even when the isola-

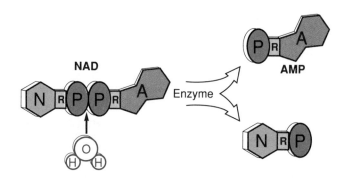

FIGURE 3-5
The coenzyme NAD (nicotinamide adenine dinucleotide) is cleaved by a water molecule, through the catalytic action of a potato enzyme, into its two components: AMP (adenosine monophosphate) and NRP (nicotinamide ribose phosphate).

tion procedure was very gentle. In trying to enhance the levels of respiration and the coupled aerobic phosphorylation, we observed that the reaction was strongly stimulated by the coenzyme NAD, closely related to NADP. Our excitement about this novel result faded when we learned that the effect could be traced to an enzymatic cleavage of NAD, using a molecule of water (Fig. 3-5).

The AMP produced by this cleavage stimulated the reaction because it served as an acceptor of two phosphates to form ATP. The stimulation was due to the inadequate amounts of acceptor AMP or ADP which were released from the disrupted kidney cells and which were required for sustaining respiration and the accumulation of ATP.

From these experiments we learned little about aerobic phosphorylation and nothing about the source of inorganic pyrophosphate. We had only a few ideas and even fewer rabbits on which to test them. Rabbits cost two dollars each, and the departmental budget limited us to only two or three a week. Lindberg and I bought our own on occasion, but our cash was also scarce. Then there was the misery of a St. Louis summer with no air conditioning. Experiments were nearly impossible for weeks because the room temperature often exceeded 37°C (98.6°F), the standard for such studies.

This marked the end of my search for the source of ATP. It had been doomed from the start because I was committed to finding discrete soluble enzymes that linked the synthesis of ATP to aerobic metabolism. They don't exist that way. Instead, these enzymes are firmly embedded in the walls of tiny intracellular organs called mitochondria. Albert Lehninger recognized this a few years later. Mitochondria are membranous sacs in which 10,000 chains of enzymes generate electric charges that mediate the flow of energy from the oxidation of a foodstuff molecule into the production of ATP. A leak or disorder in the delicate mitochondrial fabric disrupts this process.

The Warburgian allegiance to working with separated, soluble enzymes would have to be abandoned by those who would explore the operation of the organized mitochondrial enzyme system. I was not among them. Instead, I switched to a more modest project: understanding the enzyme that I had found which cleaved NAD into its AMP and niacinamide-ribose-phosphate components. Who could have foreseen that pursuit of this seemingly mundane enzyme would lead me to the synthesis of coenzymes, to the origin of inorganic pyrophosphate, and eventually to the replication of DNA?

Potatoes and the Cleavage of Coenzymes

When I returned to NIH in the fall of 1947 from pilgrimages to New York and St. Louis, my former laboratory space in the Nutrition Division (in Building 4) was occupied. Henry Sebrell, the Division Director, and Floyd Daft, the Deputy Director, showed me a tiny alcove off the diet-preparation kitchen and said it was my future laboratory. I failed to get the joke and said instantly that I would quit NIH. Chagrined, they told me that they had only been teasing me, and that space was being sought in another building.

In the meantime, Bernie Horecker gave me a bench in his laboratory in the Industrial Hygiene Division (in Building 2). Just about then, one of the organizational convulsions that frequently seized NIH threatened Bernie, and also Leon Heppel in the same division, with a transfer to Cincinnati. Fortunately, Sebrell agreed to let me start an Enzyme Section that would include Heppel and Horecker in a few laboratory rooms that had been found for us in Building 3.

The present mammoth size of NIH, covering 300 acres and employing 13,000 people, makes it difficult to recall that in 1947 there were only six small buildings. Seminars were held in a tiny rustic cottage on a site now occupied by the Clinical Center (Building 10). The external grant program that today expends $5 billion annually for 28,000 research projects around the world did not exist then. Research emphasis was still on infectious disease and was dominated by a small corps of commissioned medical officers. As members of that corps, Heppel and I might have been partying with them in the Administration Building on New Year's Eve that year. Instead, we and Horecker spent the night sneaking past the party, pushing carts of glassware and reagents from Building 2 to Building 3. Had we relied on official channels, such interdivisional transfers might have required an act of Congress.

I continued to work on the rabbit kidney enzyme that Lindberg and I had discovered in St. Louis and established that it cleaves the NAD coenzyme in the middle into its two major constituents. However, try as I might, I was unable to free the enzyme from its firm attachment to aggregates of cellular material, and so there was little hope of obtaining it in a pure form. A crude enzyme is a blunt tool and gives crude answers.

Lamb, pig, and beef kidneys were no better than rabbit as sources, nor were other tissues. I read of a potent enzyme in the brain that degraded NAD, but it did so by another route, cleaving niacinamide from the rest of the molecule. I was intrigued to learn from Sidney Colowick and Oliver Lowry of an enzyme in potatoes with an apparently similar action on the coenzyme FAD that contains the vitamin riboflavin. Sure enough, I found an NAD-splitting enzyme in blenderized potatoes and could readily extract it in a free, soluble form.

Because of considerable variation in the abundance of the enzyme among batches of Idaho, Maine, red, and sweet potatoes, I tested samples of eleven Maine varieties I obtained from a visit to the U.S. Department of Agriculture facility in Beltsville, Maryland. This formal survey identified White Rose and Teton as the best, eight times as active as Chippewa and Triumph. I returned a week later to pick up a 100-pound lot of White Rose, but found to my dismay that the activity of this batch (likely harvested from another plot at another

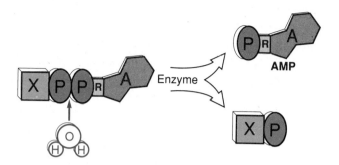

FIGURE 3-6
Any of a class of compounds resembling NAD, can be cleaved by a water
molecule between the two phosphates (at the pyrophosphate bond) to
yield AMP and the other component (a substituent linked to phosphate).
The enzyme catalyzing this cleavage is named nucleotide pyrophosphatase.

time) was as feeble as the poorest. So I simply went to a big food
market in Bethesda, asked the manager for a few potatoes from each
of a dozen nameless 100-pound sacks, and returned the next day to
buy the sack of potatoes that tested the best.

After I had purified the potato enzyme 1,000-fold from the crude
potato juice, to the point of removing most of the other enzymes, I
could show that the cleavage of NAD was matched exactly by the
appearance of its components: niacinamide ribose phosphate (NRP)
and adenine ribose phosphate (AMP; Fig. 3-5). Although the metabolic importance of this cleavage enzyme in potatoes or animal tissues was not revealed by these efforts, the purified enzyme became a
wonderful tool for exploration. To begin with, I found that the enzyme cleaved not only NAD but a whole class of compounds with
the same general structure. I called the enzyme *nucleotide pyrophosphatase* because other nucleotides could substitute for ARP and
cleavage was at the pyrophosphate (PP) bond (Fig. 3-6).

An especially interesting substrate for the enzyme was the important coenzyme NADP, related to NAD by having an extra phosphate
group. The location of this additional phosphate had remained uncertain ever since the coenzyme was discovered in 1935 because the
strong acid and other harsh chemicals needed to analyze its structure led to its destruction. I used nucleotide pyrophosphatase, a far
gentler treatment, to cleave NADP. With the help of another enzyme

from potatoes and new chromatographic methods I will describe later, I was able to show that the extra phosphate was attached to the ARP part of the NADP molecule and specifically to the second carbon of the ribose sugar (Fig. 2-9). To describe the structure of NADP in this way was a significant landmark on the map of biochemistry.

I have never met a dull enzyme. From the humble hydrolase that uses a molecule of water to split NAD to the glamorous polymerase that assembles the vast DNA chains of genes and chromosomes, the feats of enzymes are all awesome. Yet the secrets they hold are far greater. Except for a few, we know them only by trivial names and some apparent activities. For all but these few, we are ignorant of their shape, let alone their surface features and internal anatomy. We know too little of what controls their actions, speeds them up, or slows them down. We do not know enough about their family relations, how their "lives" are organized within the community of the cell and the larger universe of an organism, plant or animal.

Yet a current mood has developed that ignores or disparages enzymes. Bent on understanding the major features of cellular form and function, most biologists have turned to manipulating the DNA blueprint to discover the signals that make the enzymes which are responsible for developing the organism. The list of cancer-related genes (called *oncogenes*) has grown rapidly, and fruit flies have been created with aberrant structures. Analyzing and rearranging DNA has produced astonishing results, not least the industrial manufacture of rare hormones and vaccines. But attention to the enzymes that actually make and operate the cell has not kept pace. Without knowing and respecting enzymes, better still loving them, answers to the most basic questions of growth, development, and disease will remain beyond reach.

Synthesis of a Coenzyme

One of my greatest thrills came on a day in July 1948 when Sylvy and I returned from a two-week vacation trip to the Gaspé Peninsula in Canada. For some weeks before leaving I had been busily purifying the potato enzyme and examining a variety of its features that I would need to include in a publication describing the enzyme's properties. Upon my return, these preparative and descriptive experiments seemed dull and my enthusiasm for them had diminished.

Instead, I wondered what I might do with the NRP I could obtain as the product of the enzyme's cleavage of NAD. Might having this novel compound give me the opportunity to discover the mechanism of *synthesis* of the NAD coenzyme? Might the coenzyme be reassembled if ATP were provided (as the other component in an activated form) and the synthetic enzyme found? There was little knowledge then about how any coenzyme was assembled. Where to look for the enzyme that synthesizes it? Yeast, a rapidly growing cell easily obtained in large quantities from a local brewery, might be a good source. I also had the experience of preparing a number of different yeast extracts (juices), called *Lebedevsafts*. They were named for the Russian biochemist Alexander Lebedev, who found in 1912 that dried yeast, resuspended in water, undergoes self-digestion (autolysis) and after one or two days exudes a clear juice (*saft* in German), which can be separated from the cell debris and has a generous content of free enzymes. I also knew that my Lebedevsafts were relatively inactive in degrading NAD.

To see if I could find an enzyme that synthesized NAD, I added to the yeast extract in a small test tube the two components I thought it might need. One was a possible source of the nicotinamide part of NAD. I used a mixture in which the potato enzyme had cleaved NAD to generate NRP and was then heated in boiling water, after the NRP was formed, to destroy the enzyme. Another was ATP I had isolated from rabbit muscle. As mentioned already, ATP was a likely source of the adenosine part of NAD in an activated state. I also included a magnesium salt to put ATP into a proper form and a phosphate salt to control the acidity. Everything was kept on ice. I removed one sample immediately before any synthesis could take place and then another after the mixture had been kept at 37°C for 60 minutes; in each instance, the sample removed was immersed in boiling water for three minutes to prevent any further reaction. I then analyzed the samples for NAD in the spectrophotometer using a pure enzyme system that transfers electrons to NAD, thereby reducing it and causing it to absorb ultraviolet light.

First I analyzed the sample taken at the outset of the reaction. The spectrophotometer needle moved to a reading indicative of 0.59 micromole of NAD present in the yeast extract. (One micromole contains 600 million billion molecules.) Then I examined the sample incubated for 60 minutes. The needle moved five times as far! The value translated to 3.22 micromoles of NAD. This increment seemed reasonable because it was in the range expected from the amount of NRP that I started with. To be sure that the spectral change was due to reduction of NAD, I added an enzyme system that withdraws electrons from reduced NAD. The needle moved back to zero. A great relief!

Could this synthesis of NAD be genuine? I was excited when I went home that night, but far from believing that I had been so lucky as to discover the mechanism of synthesis of a coenzyme. I tried to think of trivial reasons that could account for these light absorption changes. The next morning, I repeated the experiment with another yeast extract and obtained similar results. The yeast extract heated to 60°C for five minutes was inactive. Later that day, with even greater relief, I found that upon omitting any one of the key reactants—the yeast extract, the NRP, or the ATP—there was no evidence for NAD synthesis. There could be no doubt that some unstable substance, presumably an enzyme, in the yeast extract reacted NRP with ATP to generate NAD.

Just what was the nature of the reaction responsible for the synthesis of NAD? Exactly how did ATP contribute to it? Relying on the crude yeast extract as the source of the synthesizing enzyme, I would find it nearly impossible to answer these questions. The extract contained many hundreds of enzymes, including some that degraded ATP and NRP. I could try to circumvent these interfering activities by finding specific inhibitors to keep them in check or try to make corrections for these unwanted reactions. However, such measures are generally clumsy and indecisive. The quickest and surest route to understanding an enzyme-catalyzed event is to purify the responsible enzyme to the point of freeing it from interfering activities.

I was very lucky in purifying the enzyme responsible for NAD synthesis. To begin with, the enzyme proved to be especially hardy during the autolysis (self-digestion) of ale yeast. Among seven different yeasts, I found one which still retained 70% of the enzyme activity even when, after autolysis for 48 hours, only 14% of all the released proteins still survived. The purity of my enzyme activity in the clear yellow yeast extract (called Fraction I) was thus increased 5-fold relative to "total" proteins, with only a modest sacrifice in yield. To the extract I then added ammonium sulfate salt cautiously. When a faint turbidity signaled that some proteins were becoming insoluble, I promptly collected that fraction by centrifugation. I

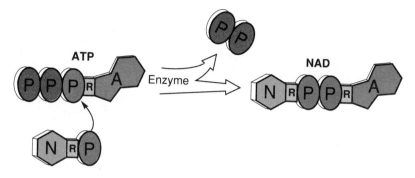

FIGURE 3-7
The yeast enzyme catalyzes NAD synthesis and the release of pyrophosphate (PP) by directing an attack by NRP on the innermost phosphate of ATP.

could hardly believe it. The barely visible pellet in the centrifuge tube, which had only 2% of the total protein, retained all the synthesizing activity. The results were equally good when I carried out the procedure on a 100-times larger scale. In this single step, I had enriched the purity of my enzyme 50-fold. I have never been as lucky since. To this preparation (called Fraction II) I again added ammonium sulfate in graded amounts and once more obtained a fraction which, this time, had 60% of the activity of Fraction II but only 6% of the protein. By now the purification had reached 2,500-fold. After several manipulations that failed, I learned that slight acidification of the preparation (to pH 4.8) precipitated 80% of the activity with only 20% of the protein. At this point, the enzyme had been enriched 10,000-fold relative to proteins readily released from a yeast cell and was approaching purity. It was now time to determine exactly how the enzyme made NAD from nicotinamide ribose phosphate (NRP) and ATP.

The results came quickly and were astonishingly clear-cut. Along with NAD, inorganic pyrophosphate (PP) was formed, the first clue to its natural origin after years of speculation. For every molecule of NRP and ATP whose disappearance I measured, the enzyme produced an equal number of NAD molecules along with the same number of PP molecules (Fig. 3-7). The reaction came to a stop in a few minutes when exactly 40% of the NRP and ATP were converted, regardless of the amount of enzyme I added; with more enzyme the endpoint was reached sooner, with less enzyme, later. The reaction

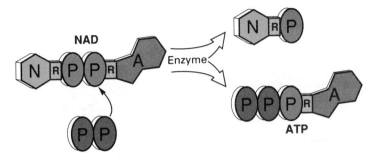

FIGURE 3-8
The same enzyme that catalyzes NAD synthesis can reverse the process, directing an attack by PP on NAD to produce NRP and ATP.

would also run backwards starting with NAD and PP, reaching the same endpoint. When I supplied NAD and PP to the enzyme to start with, 60% of each disappeared and corresponding amounts of NRP and ATP were produced. Thus, the synthesis of NAD was freely reversible (Fig. 3-8). Regardless of the direction of the reaction, whether producing NAD or splitting it, the proportions of the four reactants reached the same equilibrium state; the role of this (or any) enzyme was to hasten the attainment of the equilibrium state but not to affect the ratio of reactants at the equilibrium state.

Was this way of making NAD coenzyme unique to yeast cells? Would I find a similar enzyme in animal cells? I homogenized rat or hog liver in a blender with acetone (a solvent for fats) and dried the proteinaceous residue in air to a fine powder. (Sylvy could always tell from my clothes and breath the days I made acetone powders. This apparently innocuous procedure killed an Australian friend, Bob Morton, when the acetone vapors accidentally ignited and exploded.) Suspending the powdered liver in a solution of phosphate salts yielded extracts about as active in making NAD as those from yeast. I was greatly encouraged. Purifying the enzyme from liver proved to be far more difficult than from yeast, and I paused when the enzyme was enriched, relative to protein, by only about 100-fold. Still, this partially purified enzyme had been freed of interfering activities; it catalyzed the reversible synthesis of NAD, reaching the same endpoint (equilibrium state) as that attained with the yeast enzyme. Separated by a billion years of evolution, yeast and liver cells have the same coenzymes and, as we found later, make them the very same way.

My Rookie Year

Nineteen forty-eight, the year I set up my own biochemistry laboratory at NIH, was a great one for me. With an enzyme purified from potatoes I discovered how to cleave the complex coenzymes gently enough to leave their component halves intact. I then was able to advance our knowledge of the location of one of three phosphate groups of NADP. Cleaving NAD gave me the key to the discovery of the wondrous enzyme that makes NAD, one of my all-time favorites. I have never been more proud of an enzyme. It answered several important questions and defined a major theme in biochemistry (the generation of pyrophosphate). It also gave me instant recognition among biochemists and set me on a career devoted to the enzymes that assemble DNA, genes, and chromosomes.

I also look back on 1948 and the four years that followed as golden working years. I was at the laboratory bench all day with few interruptions. Each night, after our two young children were coaxed to bed (Tom, our second child, was born in 1948), I would sit in an easy chair with pencil and paper, and Sylvy might ask: "What are you doing?" "Thinking," I'd answer. I would be designing experiments and preparative procedures for the next day, with fall-back plans to cope with failures and disasters. At the bench, with Bill Pricer, my research assistant—able, amiable, and devoted—at my side, there were few interruptions except the daily noon break when Leon Heppel, Bernie Horecker, Herb Tabor, and I took turns giving a seminar on a paper we had read the night before.

Bill Pricer and Tony Schrecker, my first postdoctoral fellow, who was trained in chemistry at the University of Illinois, were among the earliest victims of my obsession to make every minute, every hour, and surely every day matter. Few of the many stories I hear told about me bear any resemblance to reality, but one could well be true. By this account, Bill Pricer had centrifuged a sample without balancing the tube. We all heard the tube shatter. Bill said: "Don't worry. I have more of the sample. Nothing was really lost." To which I replied: "The hour lost can never be recovered." Time has always been so precious and, having watched it so closely, I can usually guess its passage with an accuracy of a few percent—a few minutes over several hours.

Knowing the substrates and products of the NAD synthesis reaction, I could imagine how the enzyme might be working. By holding and positioning NRP and ATP in a precise way, the enzyme directs their union to form NAD and release pyrophosphate (Fig. 3-7). Conversely, the enzyme can reverse the reaction by juxtaposing NAD and pyrophosphate so that pyrophosphate displaces the NRP part of NAD to form ATP (Fig. 3-8).

With the discovery of an enzyme goes the privilege and burden of naming it. Often, the enzyme is named after the substance (substrate) upon which it acts or the product it makes. In this vein, the enzyme that directs the cleavage of the sugar maltose by a water molecule to produce two molecules of glucose was called maltase; the enzyme that brings two molecules together to form citrate (a salt of citric acid) was called citrate synthetase. But when a reaction has multiple substrates and products and is also readily reversible, the choice of a name for the enzyme becomes more difficult.

Since the enzyme that makes NAD can also cleave it, the enzyme might be called either NAD synthetase or it might be named for the reverse reaction. In the test tube (in vitro), the reverse reaction is favored slightly. In cells (in vivo), however, ATP is abundant and pyrophosphate is removed, so the reaction (NRP + ATP → NAD + PP) would be displaced to the right, toward the synthesis of NAD. Thus the name NAD synthetase would be appropriate in identifying its principal function in cells.

(I feel uneasy with having created so many words in the naming of enzymes and their actions. Language, so essential for understanding, can also be the greatest barrier to the uninitiated. I recall a seminar I gave to a group of physicists and engineers on the structure of DNA and the intricacies of its biosynthesis. Afterwards, one of them told me he had no trouble following the lecture, "But I can't understand," he said, "the difference between in vitro and in vivo.")

The mechanism for synthesis of NAD immediately suggested how another key coenzyme of cellular respiration might be made. FAD (flavin adenine dinucleotide, containing the B vitamin riboflavin) might be assembled from its halves in a similar way (Fig. 3-9). We found just such an enzyme in brewer's yeast, although the degree to which we purified it was less than that of NAD synthetase and our results were not as sharp. Part of our problem was the low abundance of the FAD coenzyme in cells and hence the meager amount of riboflavin phosphate (less than a milligram) we could get from FAD by cleavage with our potato enzyme. Some time later I learned that grams of riboflavin phosphate could be had for the asking from a

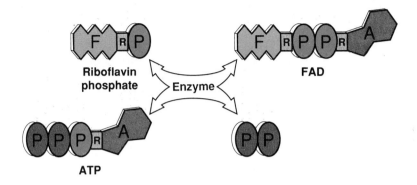

FIGURE 3-9
Synthesis of the coenzyme FAD (flavin adenine dinucleotide) from
riboflavin phosphate and ATP follows the same pattern observed in the
synthesis of NAD (Fig. 3-7). As the double arrows indicate, this reaction
is also reversible.

pharmaceutical company that made it chemically as part of a soluble
vitamin formula for infant use.

In the excitement of having found the mechanisms for synthesis of
two of the major respiratory coenzymes, NAD and FAD, I went after
that of the third coenzyme, NADP. It differs from NAD in having the
extra phosphate group which I had found some months earlier was
attached to its ARP half. I found an enzyme in brewer's yeast that
made NADP (Fig. 2-9) by transferring the terminal phosphate of ATP
to NAD.

In the published account of how the enzyme was purified from
yeast, the first entry was the value for the enzyme activity in the
ammonium sulfate fraction prepared from the yeast extract. Since I
could detect no activity at all in the crude extracts, the value was
zero. Michael Doudoroff, a biochemist and microbiologist at the University of California, Berkeley, told me years later that he admired
my courage and ingenuity in proceeding with the purification of an
enzyme from a source which appeared to lack it. I had to tell him the
truth. I first looked for the enzyme in ammonium sulfate fractions I
had prepared for other purposes. Having found it in one such fraction, I went back to the extract from which it was prepared, only to
discover that in this cruder source the activity was masked by an
inhibitor. I had no interest in the inhibitor except to avoid it. It was a

ship that I let pass in the night. But I never forgot the lesson. Thirty-three years later, after many years of unsuccessful attempts to find a cell extract active in starting a cycle of DNA replication, the key to success turned out to be proceeding with the preparation of an ammonium sulfate fraction from an inert extract.

In the spring of 1949 I entered the United States Marine Hospital of the U.S. Public Health Service in Baltimore for repair of two inguinal hernias. I brought a briefcase bulging with data for the papers that would describe the work of my previous rookie year: "Nucleotide Pyrophosphatase," "Reversible Enzymatic Synthesis of NAD and Inorganic Pyrophosphate," "Reversible Enzymatic Synthesis of FAD," and "Enzymatic Synthesis of NADP." When I was shown my room, I snapped at the nurse: "There's no desk here." How would I spend the rest of that day and the two or three days after surgery without a place to spread my notebooks and write the papers? The nurse laconically asked: "Did you see that uniform on the cart in the hall? It belonged to a young medical officer who didn't survive his 'minor surgery.' " I didn't die on the operating table, but I also didn't miss having a desk during the painful postoperative days.

The four manuscripts were finally completed that summer and appeared in the February 1950 issue of the *Journal of Biological Chemistry.* How coenzymes and inorganic pyrophosphate are made and the novel chemistry Nature uses to make them interested a considerable fraction of what was then a small body of biochemists. This research earned me the annual award in enzymology (then the Paul-Lewis and now the Pfizer Award) and attracted applications for postdoctoral training from some of the most gifted young biochemists in the country. Among them in 1952 were four future stars: Bruce Ames, Paul Berg, Edward Korn, and the late Gordon Tomkins. Had Korn eventually joined me, as Berg did, we might have produced a paper by Korn, Berg, and Kornberg.

In the ensuing years, the mechanism of biosynthesis of the relatively simple coenzymes that I had discovered was rediscovered again and again in the biosynthesis of the much larger "macro" molecules of cells and tissues. Four classes of these macromolecules—proteins, lipids, carbohydrates, and nucleic acids—make up the structure and machinery of living matter. They are polymers, chains of tens to many thousands of building blocks. The building blocks of a protein are amino acids. Each amino acid first reacts with ATP. Pyrophosphate (PP) is released and the amino acid

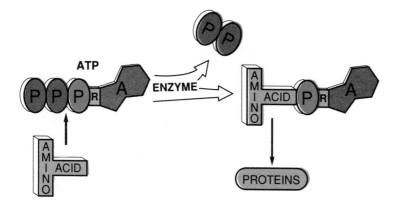

FIGURE 3-10
ATP activation of amino acids, the building blocks of proteins, takes
place by a mechanism similar to that of coenzyme synthesis.

thus activated is subsequently incorporated into a particular protein
sequence (Fig. 3-10). To make fats and steroids, acetic acid units are
activated by ATP in a similar way, also generating PP. The syntheses
of carbohydrates and the phospholipids of membranes employ a
slightly different wrinkle; their building blocks react with analogues
of ATP in which adenine is replaced by uracil or cytosine, but the
reaction mechanism, with the release of PP, follows the same pattern
as that of amino acids and coenzymes. Finally, the enzymes responsible for the synthesis of nucleic acid chains (DNA and RNA) assemble the building blocks in a manner that once again resembles that of
coenzyme biosynthesis, with the release of PP.

One Gene, One Enzyme

That a disease could be due to the absence or malfunction of a single
enzyme was first proposed by Sir Archibald Garrod at the beginning
of the twentieth century. He had patients whose urine turned black
on exposure to air because it contained massive amounts of homogentisic acid, a metabolic product of the amino acid tyrosine.
This aberration, called alkaptonuria, while rare in the general popu-

lation, was common among siblings of the patient. It was also more prevalent in the progeny of blood-related parents. Garrod observed a similar familial history in people who excreted excessive amounts of the amino acid cystine (cystinuria) or the sugar pentose (pentosuria).

Apprised of the recently rediscovered findings of Gregor Mendel on the inheritance of traits, Garrod developed the concept that homogentisic acid, cystine, and pentose were normally produced in the body but could not be processed further in these patients because of a hereditary lack of the specific enzyme needed for the job. Albinos, Garrod reasoned, represented another such inborn error of metabolism, one in which the formation of skin pigment was blocked at some stage because of the deficiency of an enzyme.

Not until 1909 was the term *gene* used for the hereditary determinant of a particular characteristic. Then thirty years elapsed before Garrod's hypothesis was recalled and incorporated into the correct picture. George W. Beadle and Edward L. Tatum, examining the results of mutations in the red bread mold *Neurospora crassa*, postulated that one gene encodes the information to produce one enzyme. They had exposed the microbial culture to x-rays to damage the DNA. At sufficient dosage, fewer than one in a million cells survived to grow on any medium. Of the survivors, a small percentage failed to grow on the simple medium of sugar, mineral salts, and the vitamin biotin, which is all that is needed to sustain normal *Neurospora*. These mutant strains did grow when provided with a medium enriched with yeast extract, which contains the various other vitamins, amino acids, and nucleic acid constituents.

In most instances, Beadle and Tatum could identify a single nutrient as the substance needed to sustain the mutant microbe. Evidently the microbe had sustained x-ray damage to the gene responsible for the enzyme that normally makes this nutrient for the organism. From these and earlier studies on eye-color mutants of the fruit fly, *Drosophila*, Beadle and Tatum could extrapolate what had not been sufficiently clear to Garrod, namely:

(1) that all biochemical processes in all organisms are under the control of the genes,

(2) that each of these processes can be resolved into a series of individual stepwise reactions, each under the control of a single different gene,

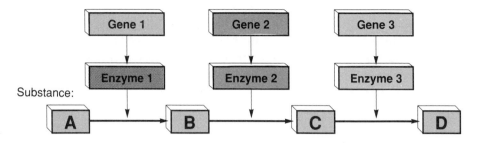

FIGURE 3-11
A metabolic step (cellular chemical event) is catalyzed by a distinctive enzyme specified by a single gene.

(3) that alteration of one gene results in a deficiency of only one enzyme responsible for directing a single chemical reaction (Fig. 3-11).

The metabolic consequences of a defect in a single type of enzyme or protein, among the ten thousand or more that run our body chemistry, can in many instances be anticipated from what we know of its functions in metabolism. Lack of Protein VIII in the blood-clotting chain results in hemophilia; alterations in hemoglobin produce a variety of anemias. I was therefore startled to learn from a scientific report in 1967 that a grotesque and fatal illness resulted from a defect in hypoxanthine guanine phosphoribosyl transferase (HGPRTase). Having discovered this enzyme, I thought I knew it rather well and had relegated it to an auxiliary and dispensable role in nucleic acid metabolism.

One of the patients referred to in that report was a boy of four and a half years, called M.W. He was admitted to the Harriet Lane Home of Johns Hopkins Hospital in Baltimore because of bloody urine, fever, and vomiting. Apparently normal at birth, he had failed to develop properly. When seen at the clinic at eight months of age, he was markedly spastic and had irregular, involuntary movements of his arms and legs. Now he appeared to have advanced cerebral palsy (paralysis). He could not stand or sit by himself. He spoke little. Most unusual was a compulsive biting of his fingers and lips. He seemed terrified and screamed as if in pain while biting his fingers and was calmed only when securely restrained.

His urine had copious amounts of a sharp, crystalline material. The substance was a salt of uric acid, the end waste product of the metabolism of constituents of RNA and DNA, the genetic materials. The uric acid concentration in his blood was very high (called hyperuricemia), far beyond the level at which it forms crystalline deposits in tissues and joints.

Gout, in which hyperuricemia leads to such needle-like deposits, especially in the joints and kidneys, is a painful and disabling disease, but one that is practically never seen in children. Nor are neurologic changes observed in gouty patients, unless something can be gleaned from the behavior patterns of such alleged gout victims as Ben Jonson, Isaac Newton, Cotton Mather, Benjamin Franklin, Samuel Johnson, and Charles Darwin.

The twin anomalies of hyperuricemia in a child and compulsive self-destructive biting, as well as cerebral palsy and spasticity, were also observed in M.W.'s eight-year-old brother. He too could not sit or crawl and appeared mentally retarded. He had begun biting his hands at the age of four and had partially amputated the tip of one finger and completely chewed away his lower lip to the point where it was beyond the reach of his front teeth.

When Michael Lesch, a medical student, and Dr. William L. Nyhan reported these cases in the *American Journal of Medicine* in 1964, their survey of the literature since 1823 had turned up but fifteen cases of hyperuricemia or gout in patients under 10 years of age; and of these, only two showed abnormal neurologic symptoms. Alerted by the 1964 report and assisted by greatly improved and routine analyses for uric acid, physicians soon reported well over a hundred cases resembling those of the Baltimore children. Careful studies of these cases of what would become known as Lesch-Nyhan syndrome disclosed the disease to be hereditary in origin, due to a defective gene located on the X chromosome, one of the two sex chromosomes. Like hemophilia, it was transmitted from mother to son. The altered gene produces a defective HGPRTase enzyme in place of the normal one, whose function is to salvage breakdown products of RNA and DNA before they are further altered and irretrievably lost as uric acid. While it is now clear why the lack of HRPRTase results in hyperuricemia and gout, the cause of the nervous system disorder remains a mystery.

We now know that Lesch-Nyhan syndrome, a mutation which changes just one of the 654 bases in the DNA of the HGPRTase gene, and thus one of the 218 amino acids in the salvage enzyme it encodes, can cause a young child to have gout, severe palsy, obsessive self-mutilating behavior, and little hope of reaching adulthood. By 1982 some 1,400 diseases, each due to a defect in a single gene, had been described in the medical literature. Their combined frequency in the population is about 1 per 100 live births, and they account for about 5 percent of hospital admissions among children. For most of these diseases (for example, cystic fibrosis, Huntington's chorea), the biochemical basis is still unknown. For approximately 200 of them, the disease is known to be due to a deficiency or malfunction of a single enzyme.

Chapter 4

Bless the
Little Beasties

That DNA is somehow associated with heredity is now common knowledge among the flight attendants and cocktail-party-goers whom I poll on occasion. The popularity of recombinant DNA technology as a financial investment may have contributed to this understanding. Nevertheless, I have not found any lay person who could clearly state the basic, dual functions of DNA: to serve as a stable template for replicating itself, and to provide detailed information for making the cell's enzymes.

Most people know that the enormously intricate organism that emerges from the union of the sperm and egg cells of mice—an organism made up of trillions of cells—is invariably a mouse. Some people also realize that chromosomes in the nucleus of each cell contain the hereditary information that dictates every subtle detail of an individual—hair color, nose shape, and so on. Usually, the lay person does not understand that the thousands of genes in a chromosome also send individual instructions for making every minute detail of the cell—that the chromosome functions as a construction manual for the cell.

DNA is the chemical language in which this construction manual is written. It is an atomic language, the ultimate in miniaturization. I like the analogy first suggested by Richard Feynman, the eminent physicist. In 1960, in an article entitled "There's Plenty of Room at the Bottom," he offered a prize of $1,000 to anyone who could miniaturize the *Encyclopaedia Britannica*, all 24 volumes, to the size of a pinhead. The prize was finally won in 1986 by an electrical engineering student at Stanford working with ultrasmall electronic devices. This feat of microminiaturization of language may seem fantastic, yet the dot at the end of a sentence in a pinhead-sized *Encyclopaedia Britannica* covers an area which can include a chain of a thousand distinctive atoms. It is on this unimaginably small scale that the language of DNA is written, a language which evolved billions of years ago in tiny one-celled creatures.

Much of what we know about DNA and genes has been learned from the smallest and simplest of life forms, the viruses. The chromosome of the smallest virus contains only a few genes, compared with the many thousands found in a bacterial or animal cell. Among the most instructive of the viruses have been those that infect bacteria. *Escherichia coli*, the common intestinal bacterium and favored microbial guinea pig, plays host to a large variety of viruses. One of the intensively studied of these bacterial viruses, called *phages* (Greek *phagein*, to eat), is called T2.

As seen in the electron microscope (Fig. 4-1), the T2 phage is a complex creature. It has a head with thirty facets, a collar piece around a neck, a tail, and a tail plate into which are inserted exactly six fibers. These structures—the head, collar, tail, and fibers—are made up of proteins. Packed into the head is the virus's DNA, wound tightly as a coil. The tail fibers recognize a particular spot on the surface of *E. coli*. (Fig. 4-2). When properly anchored, the tail piece burrows a hole in the bacterial wall and injects the phage's DNA into the cell. The DNA, and DNA alone, is responsible for producing new virus particles; the protein structure—the elaborate head, collar, tail, and fibers—serve only to protect the virus's DNA in nature and to function as a syringe for injecting the DNA into the host cell.

Injection of the DNA takes only a second. Immediately, in some still mysterious way, the cell's own DNA is immobilized and the viral DNA takes over the cellular machinery, forcing it to make hundreds of copies of the viral DNA (Fig. 4-3). Each copy (chromosome) is enclosed in a head, to which the other parts become attached. The construction of the head, neck, and tail are also under control of the

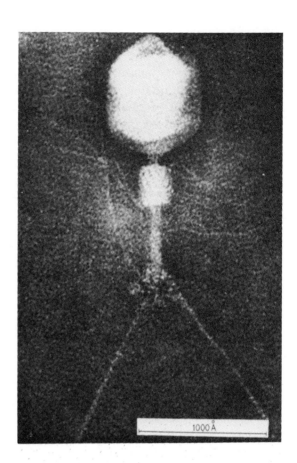

FIGURE 4-1
A bacterial virus (T2 phage) seen in the electron microscope. 1,000 Å is one-ten-thousandth of a millimeter.

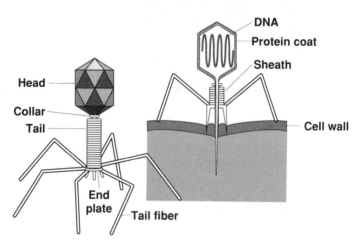

FIGURE 4-2
Anatomy of the T2 phage. After the tail fibers attach to the surface of a bacterial cell, the tail contracts, causing the DNA packed in the head to be injected into the cell in a syringe-like fashion.

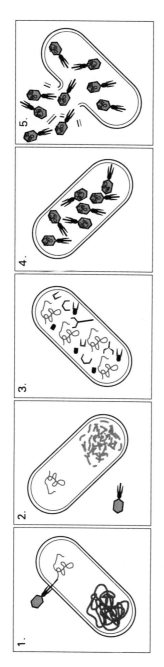

FIGURE 4-3

Cycle of phage infection, completed in about twenty minutes. (1) Entry of the phage DNA (chromosome) triggers (2) destruction of the cell's DNA. (3) Under the direction of the phage chromosome, the constituents of the cell's DNA are used as building blocks to assemble several hundred copies of the phage chromosome, and proteins within the cell are assembled into viral heads, necks, tails, and so on. (4) The various parts come together to form infectious phages, which are then released from the cell (5).

viral genes. Within twenty minutes several hundred new, infective viruses accumulate in each cell (Fig. 4-4) and are then released.

When the head of a virus is artificially ruptured, the tightly packed DNA streams out and can be seen as an extremely long fiber (Fig. 4-5). The DNA fiber (the *chromosome*) of T2 is fifty microns (50 millionths of a meter) long. Even magnified a thousand-fold, it would still be barely visible to the naked eye. Yet it contains more than a hundred distinct genes arranged in a precise linear sequence. Each *gene* is a stretch of DNA that includes a string of about one thousand DNA components, called *nucleotides*. The gene carries a message spelled out by the sequence of these nucleotides. This message is first transcribed into RNA and then translated, using the genetic code, into a corresponding sequence of amino acids. These amino acids join together to make a particular protein. One such gene in the T2 virus would specify the head protein, another the tail protein, and so on. (In just the same way, one of the human chromosomes contains a single gene that defines the precise structure of the hormone insulin, for example.)

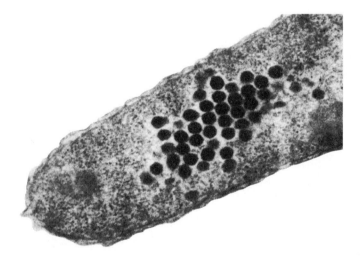

FIGURE 4-4
An electron microscopic view of a section through an infected bacterial cell showing arrays of phage heads (with chromosomes inside) awaiting attachment of tails and tail fibers.

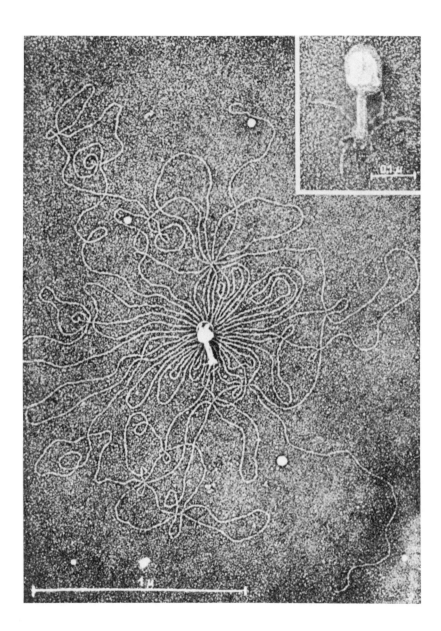

FIGURE 4-5
The chromosome (DNA) extruded from the head of a T2 phage is seen in the electron microscope as a continuous strand about fifty-one thousandths of a millimeter.

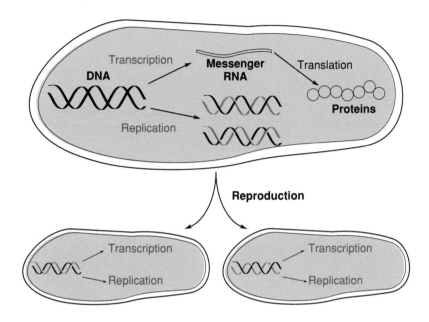

FIGURE 4-6
Dual roles of DNA: *transcription* (via RNA) to produce proteins, the agents that determine every detail of cellular form and function; and *replication* to provide two identical copies that will be shared by the two daughter cells upon reproduction.

To repeat, then, we can regard DNA, whether in a virus or a person, as having two roles (Fig. 4-6): (1) to supply information to produce the proteins which provide the distinctive machinery and structure of each cell; and (2) to serve as a template for replication in order that the DNA of two daughter cells will be identical to that of their parent.

In the mid-fifties this information flow from DNA → RNA → protein came to be regarded as "the central dogma" of molecular biology. Then in 1970 came the discovery that tumor viruses (and more recently the AIDS virus), which are made of RNA rather than DNA, are first transcribed into DNA, and only later back to RNA. Thus: DNA RNA → protein. Once the RNA viral chromosome has been restated in DNA language, it can insinuate itself into one of the host's own chromosomes. The consequences of this viral residence in the host chromosome are often profound. While in place, the viral DNA may so distort the orderly expression of the cellular genes as to produce cancer. When, on certain occasions, the viral DNA in the chromosome is transcribed back into its natural RNA state and enveloped in protein coats, these copies of the virus can emerge from the host cell to infect other cells.

The publicity attendant upon the discovery of reverse transcription of RNA into DNA, headlined "Central Dogma Overthrown," nettled Francis Crick, the scientist most clearly associated with the precept (and one of the codiscoverers of the double-helical structure of DNA). He objected that the central dogma never precluded transcription back and forth between DNA and RNA; the only thing precluded was information flow from proteins back to nucleic acids. It amused me that Crick, who had a few months earlier advised me to ignore distorted interpretations of my work by *Nature*, was now so upset by this journal's erroneous version of his own.

The Building Blocks of DNA

After infecting a cell, some phages direct the synthesis of their DNA at a rate ten times faster than the uninfected bacterium can synthesize its own DNA. In the early 1950s, no one had the foggiest idea how cells make DNA, let alone how a tiny virus could preempt the cellular machinery and raise its productivity of DNA many-fold. Nor was there any knowledge of the form of the building blocks that cells or viruses used to make their DNA. I became interested in this problem in 1950 while still at the National Institutes of Health. Feeling emboldened by my recent successes in finding enzymes responsible for the biosynthesis of several coenzymes, I hoped eventually to find an enzyme that assembles the building blocks of DNA and RNA into nucleic acid chains.

What indeed are the building blocks of DNA? A nucleic acid (RNA or DNA) can be seen as having a backbone of perfectly alternating phosphate (P) and sugar groups (designated R for the sugar ribose). Each phosphate in the backbone is attached to carbon number three of the sugar on one side and carbon number five on the other (Fig. 4-7). (In DNA, unlike RNA, the ribose lacks an oxygen on its second carbon and is called 2-deoxyribose.) Attached to each sugar at car-

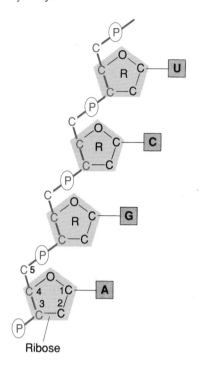

FIGURE 4-7
The DNA backbone is made up of alternating ribose and phosphate units—phosphate attached to
carbon 5 of one ribose and carbon 3 of the next in
the chain.

bon number one is one of the four bases or "letters" of the RNA or
DNA language: adenine (A), guanine (G), cytosine (C), and uracil (U)
in RNA or thymine (T) in DNA.

Examination of the structure of the nucleic acid chain did not
make it at all obvious to me in 1950 what its building blocks might
be. Was the backbone assembled first and were the letters attached
later? Was each link added to the chain as a whole nucleotide (that
is, A-sugar-phosphate, and so on)? If so, was the phosphate in each
component nucleotide initially attached to carbon number three or
five, or to either one randomly, or in a cyclic form to both (Fig. 4-8)?
In trying to anticipate what the building blocks might be, I was
influenced by what I had learned from the biosynthesis of coenzymes. I also felt that in searching for the form of the nucleotide that

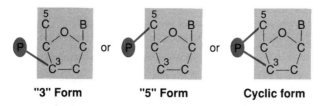

"3" Form   "5" Form   Cyclic form

FIGURE 4-8
It was once thought that the nucleotide units used in assembly of the DNA backbone might be in any one of three forms. B denotes a purine or pyridimine base.

might serve as a building block for RNA and DNA, it would help to know how a nucleotide itself is built from simpler molecules (amino acids and carbon dioxide), and thus what its nascent form might be.

When I explained in Chapter 3 how cells make the NAD coenzyme from nicotinamide ribose phosphate (NRP) and ATP (adenosine triphosphate), the reader might well have asked: "Where do the NRP and ATP come from?" (Most molecular biologists today would simply assume they came from a chemical company catalogue.) I would have answered that the nicotinamide part of NRP is a vitamin for animals, obtained by eating vegetables (or other animals which have eaten vegetables) that make it from carbon dioxide (CO_2) in the air and nitrogen in the soil. The ribose sugar and phosphate are then attached to the vitamin inside the animal's cells. As for ATP, I would have to start way back to describe how cells make the adenine (A) for ATP and for the nucleic acids (RNA and DNA).

At the end of World War II John Buchanan, a young biochemist (then at the University of Pennsylvania, and later at the Massachusetts Institute of Technology), had obtained the first clues to the small molecules that cells use to build adenine. He resorted to a relatively new technique that would prove to be the most powerful in biochemistry, the use of isotopes to trace the fate of molecules. This isotopic tracer technique actually originated as a failed experiment shortly after the turn of the century.

In 1912 Georg von Hevesy (1885–1966), a Hungarian physicist working in Lord Rutherford's laboratory in Manchester, England, was finding it impossible to separate two products of radium decay. One intermediate product, radium D, was still radioactive; the other, the terminal product, possessed no radioactivity. In their chemical properties, both radium-decay products resembled lead, which is radioactively inert. Fortunately for Hevesy, Frederick Soddy (1877–1956), in the same laboratory, was finding that elements of different mass could have identical chemical properties. He called them *isotopes*. "They are chemically identical, and save only as regards the relatively few physical properties which depend upon atomic mass directly, physically identical also." Radium D, Hevesy concluded, is an isotope of lead, the terminal product of radium decay. As Soddy expressed it simply: "Isotopes are alike on the surface but unlike inside."

The chemical properties of an atom or element are determined by electrons, the very light particles with a negative electrical charge that swarm about the positively charged nucleus. The mass of an atom, however, is determined by the number of heavy particles in its compact nuclear core. For example, the chemical behavior of hydrogen, the simplest element, depends on the one electron circling around it, while its mass is determined by the one proton (a particle with a positive charge) in its nucleus (Fig. 4-9). There are two other isotopes of hydrogen: deuterium and tritium. Deuterium, with one additional uncharged particle (a neutron) in its nucleus, has twice the mass of hydrogen. Yet, the chemical behavior of "heavy" water, which is water made up of deuterium and oxygen rather than hydrogen and oxygen, is much like that of ordinary water. Even tritium,

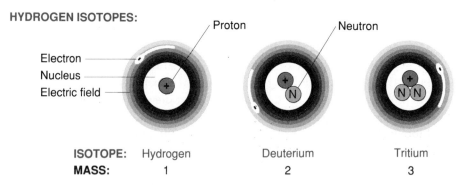

HYDROGEN ISOTOPES:

Proton   Neutron

Electron
Nucleus
Electric field

ISOTOPE: Hydrogen   Deuterium   Tritium
MASS: 1   2   3

FIGURE 4-9
Hydrogen isotopes are characterized by the number of neutrons (heavy, electrically neutral particles) in the nucleus. All hydrogen isotopes possess only one electron, and thus they all have the same chemical properties.

with two neutrons added to the hydrogen nucleus and a mass three times that of hydrogen, still forms water and resembles hydrogen chemically, because there is but one electron balancing its single nuclear proton.

Hevesy's inability to separate the lead isotopes chemically made him realize that a radioactive isotope could be an extraordinary device to trace the chemical behavior of an element because its behavior is identical to the nonradioactive form. The only difference is that the isotope, being radioactive, can be easily traced. Unweighably small quantities of the tracer would suffice because of the precision and sensitivity of radioactivity measurements. The decay of a single atom can be detected in a scintillation counter; to appreciate this sensitivity, one needs only to realize that it takes a billion billion atoms to make one milligram of salt. Technical advances since Hevesy's day have made it possible to use stable (nonradioactive) isotopes, of which deuterium is an example; tritium is a radioactive isotope. When put in place of a particular hydrogen in a sugar molecule, either isotope serves as a faithful reporter of its own fate, without perturbing the metabolism of the compound in which it resides. Deuterium, for example, is found everywhere in nature—in plants, animals, and sea water—in the same abundance relative to hydrogen: one part in five thousand. The striking constancy of this ratio attests to the absence of any significant metabolic discrimination between hydrogen and deuterium.

To trace the source of adenine, Buchanan fed pigeons certain amino acids and one-carbon compounds (such as carbon dioxide and formic acid) tagged with isotopes of carbon or nitrogen. From the bird droppings, he isolated uric acid, the excreted form of adenine, which retains the skeleton of five carbon and four nitrogen atoms (Fig. 4-10). By determining which of the carbon and nitrogen atoms in the uric acid was isotopically labeled, he inferred that the simplest amino acid, glycine, contributed all its carbon and nitrogen atoms to uric acid; carbon dioxide and formic acid were the sources of the other carbons.

That was about as far as Buchanan could go with this experimental approach. Information about a cellular metabolic process obtained by feeding tagged nutrients to a plant or animal and analyzing what it makes and discards is at best incomplete and inconclusive. It is like trying to determine the details of one of many operations in a factory without being able to enter the building. Snooping into the

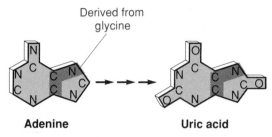

Adenine   **Uric acid**

FIGURE 4-10
Adenine is synthesized in the cell from simpler molecules, including the
amino acid glycine. This fact was learned by analysis of uric acid, the
end-product of adenine metabolism (oxidation) in birds and humans.

raw materials at the receiving dock, examining the products, and
sifting through the wastes in the trash bins is a crude process at best.
There is no alternative, ultimately, in dealing with this metaphorical
factory but to hunt among the objects on the inside for the machines
responsible for the operation of interest.

The machines that perform operations within the cell are the enzymes. In disrupted cells, one must search among the thousands of
enzymes for the particular few that make adenine from glycine, carbon dioxide, ammonia, formic acid, and other nutrients. This is what
Buchanan and G. Robert Greenberg (then at Case Western Reserve
University Medical School in Cleveland) were doing with subcellu-
lar particles of liver homogenized in a blender or with clear cell juice
extracted from defatted liver powder.

Inasmuch as Buchanan and Greenberg were looking into the assembly of the purine (A and G) nucleotides, I decided in 1950 to go
after the biosynthesis of the pyrimidine (C, U, and T) nucleotides. I
was also turning sharply to microbial cells, convinced of the superior terrain they provided for exploring basic biochemical mechanisms and of the universality of these mechanisms in nature.

Enrichment Cultures

The capacity of soil microbes to degrade virtually all components of
plants and animals was first impressed on me in the course of my
attempts to learn how the three pyrimidine precursors of RNA and

DNA were themselves manufactured in the cell. I looked for some sign of their synthesis or degradation in the juices extracted from liver or yeast cells but found none. At this juncture, Osamu Hayaishi, later to become the doyen of biochemistry in Japan, had come from Osaka, and was working with me at NIH as a postdoctoral fellow. He did this simple but impressive experiment. He scraped a little soil ("Don't call it dirt") from a car tire in the parking lot and suspended it in flasks containing a salt solution, each with a different precursor. I was astonished the next morning to find that the precursor had vanished from the solution, which was now cloudy with swarms of bacteria. From them, we cultured pure colonies (*clones*) of a bacillus with the special capacity to use cytosine, uracil, or thymine as the sole source of energy, carbon, and nitrogen for growth and reproduction. The juices we extracted from these bacteria were effective in converting cytosine, uracil, or thymine into barbiturates, but this degradative pathway seemed unpromising for clues to the synthesis of these nucleic acid precursors.

By this enrichment-culture technique, we were exploiting the fact that soil, with a billion microbes per gram, contains thousands of species, each uniquely endowed to recognize and respond to some one or several of the great variety of plant and animal substances. When that substance appears in the soil, microbial species able to make use of it will multiply and dominate the population of that sample.

A year later, in 1951, I wanted to explore the synthesis of orotic acid, closely related to uracil and considered at that time to be a more likely pyrimidine precursor than uracil (see Chapter 5). I had made no headway with extracts of animal or yeast cells, nor could I obtain an enrichment culture. My close friends and colleagues Terry and Earl Stadtman urged me to get the expert guidance of H. A. Barker, their revered professor at the University of California, Berkeley. This would also give me a good reason to escape from Washington that summer and visit California for the first time.

Trained as a chemist at Stanford University, Barker had, during postdoctoral work, been influenced by the great microbiologist C. B. van Niel in Pacific Grove, California, and then van Niel's teacher, A. J. Kluyver in Delft (The Netherlands). Barker had made outstanding discoveries about the metabolism of fats and nucleic acids through the skillful use of microbes obtained by enrichment culture. He had his greatest success with soil bacteria that thrived only in the absence of oxygen. Degradation of a nutrient substance by such bacteria does not proceed as far as with aerobic organisms, but the limited fermentations are usually more vigorous and biochemically revealing.

Barker took me to his favorite brackish pond near the San Francisco Bay shore, where I scooped up a few spoonfuls of mud. These I put into small glass-stoppered bottles with a solution of orotic acid, filled to overflowing, leaving no air space. With care to keep the cultures free of oxygen, I was able to isolate a novel microbe especially adept at converting orotic acid to smaller units which later proved to be valid precursors in the synthesis of the nucleic acids. (Barker later found this microbe to be adept, like yeast, at fermenting glucose. With his naming it *Zymobacterium oroticum* instead of *Zymobacterium kornbergii* went my best chance for immortality.)

Were orotic acid from decaying plants and animals not readily degraded by soil microbes, it would, after some millions of years, accumulate as mountainous deposits. In fact, there are such deposits containing the nucleic acid component guanine on coasts and islands frequented by sea fowl. In the partially decomposed excrement, called guano, the guanine has somehow eluded microbial action and is mined for its value as fertilizer.

The Invisible World of Tiny Animals

A few years later, when I tried to describe my work on vitamins and enzymes to my three little boys, Rog, Tom, and Ken (who was born in 1950), I found it difficult to explain what vitamins and enzymes are and why we need them. I could have said vitamins are substances the body needs to grow but cannot make for itself. I could have described enzymes as tiny machines that make and break molecules. But that would have been dull and vague. I could find no analogies to substitute for chemical formulae and reactions essential for explaining vitamin structure and how they contribute to enzyme function in human metabolism. So I shifted attention to earlier events in the history of medicine and biochemistry, a history that I was then retracing in the erratic course of my own career.

Stories about vitamins and the chemistry of metabolism would have to wait until I had told my boys a series of "germ stories"—

tales about the tiny "animals" that had entered the bodies of people, real and mythical, and made them gravely ill. These bacteria, fungi, and protozoa had euphonious names and bizarre life cycles and caused exotic symptoms. You could see these little creatures in the microscope and watch them swim about. They ate, digested, and excreted; they grew and multiplied.

Some of the stories were about the hunters who fearlessly tracked and first discovered the germ that caused a particular illness; some were anecdotes of my own encounters in the hospital or on a ship at sea. But there were also less scary stories about the good germs that form the massive colonies we house in our intestines and whose rents we collect in digestive services and the vitamins they produce for us.

Crusades to wipe out smallpox virus and other disease-causing microbes from the face of the earth have engendered the widespread feeling among most people that the only good germs are dead germs. Yet of the untold number of microbial species alive today, only a few cause disease. For four-fifths of the three billion years that life has been on this planet, microbes performed the evolutionary experiments that gave rise to the animals, plants, fungi, and specialized bacteria that now surround us.

"Yes, I'll tell you a *bad* germ story, but the next one must be about *good* germs." My children (and, later, my grandchildren), not unlike adults, have been more excited by microbial crimes and disasters than by the feats of soil microbes, cheese bacteria, and yeast cells. The common baker's or brewer's yeast (*Saccharomyces cerevisiae*) (Fig. 4-11) has been good to man from the earliest of times and is the hero of many good-germ stories. Wine, beer, and bread making,

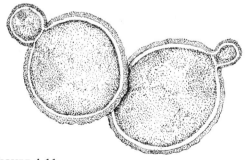

FIGURE 4-11
Yeast cells, shown here reproducing by budding,
are about five times the length of bacteria.

practiced by the ancient Hebrews and Egyptians, depend on the conversion of sugar to alcohol in beverages and the generation of gas (carbon dioxide) to impart the texture to leavened bread. Microbes are also essential to the production of all cheeses (*Penicillium* species ripen Roquefort and Camembert), and fermented milks (yogurt, sour cream, buttermilk) and are responsible for the production of food flavors, vitamins, industrial solvents (acetone, butyl alcohol), and antibiotics (many millions of pounds of penicillin and tetracyclines).

Bacteria that form nodules on the roots of leguminous plants such as clover, alfalfa, and soybeans convert the nitrogen in air to an ammonia form that these plants can use, thereby avoiding the need for nitrogen fertilizer. The complex assortments of microbes in the rumen (first stomach) of cows, sheep, deer, camels, goats, and monkeys break down the cellulose of wood, leaves, and grasses to forms the microbes can assimilate for their own growth and that of their host. (The cow's rumen, a 100-liter tank fed daily by 100 liters of saliva, contains an ecologic community of microbes whose intricate operations and balance still defy adequate description.) When microbes are harnessed to the needs of humanity, they can produce cheap high-protein foods, salvage the oil in spent wells (generally as much as half the initial content), remove toxic wastes and pollutants, convert sewage to useful fuels, and serve as factories for genetic engineering.

One relaxed summer my supply of stories was enormously enriched with the heroics of the soil bacteria, which play an essential role in maintaining life on this planet. Humans feed on vegetation or on other animals that subsist on vegetation. But what do vegetable species subsist on? They use solar energy to convert carbon dioxide to carbohydrates (sugar, starch, cellulose), fats, and proteins. Were all dead vegetation and animal matter deposited in soil to remain undecomposed, the supply of atmospheric carbon dioxide might be exhausted in a few decades. Then all life would cease. Obviously, a means must exist for cycling the carbon dioxide back into the air by decomposing the carbohydrates, fats, and proteins of dead plants and animals. This is what the soil bacteria accomplish.

In those rare instances in which carbon dioxide fails to be recycled back into the air, the consequences are impressive. The calcareous (calcium carbonate) skeletons of coral that form the charted reefs and atolls exceed in mass all man-made construction on earth. Deposits of fossil fuels—coal, oil, and natural gas—owe their origins to plants and protozoa that remained inaccessible or resistant to microbial decomposition.

These stories were told during the summer of 1953, when we were in Pacific Grove on the Monterey Peninsula in California. The five of us were living in a one-bedroom Chinese pagoda transplanted by a retired army officer from the Philippines to the sands on the northern boundary of the Asilomar Conference Grounds. There were no other houses on Pico Avenue then to spoil the lovely white dunes that stretched down to the rich tidepools at the shore and the ocean beyond. We were in Pacific Grove because I was enrolled in the celebrated microbiology course that Cornelis (Kees) B. van Niel (1899–1985) gave once a year to twelve students at the Hopkins Marine Station of Stanford University.

We had moved in January of that year from Bethesda to St. Louis, where I had become Chairman of the Department of Microbiology at Washington University School of Medicine. The name of the department had been changed from Bacteriology and Immunology, but the exclusive curricular concern, as in all such medical school departments, was with the diagnosis and course of microbial diseases and the body's immune responses. My metamorphosis from physician to biochemist and my growing awareness of genetics had convinced me that instruction of medical students in the basic biochemistry and genetics of bacteria, viruses, and parasites would be more valuable than exclusive attention to the latest techniques in culturing and staining each of the many pathogenic microbes. As chairman of a department that would try to teach these aspects of general microbiology, it seemed to me that I should take some formal instruction in the subject. I also sought refuge for myself and my family from the steamy heat of a St. Louis summer.

Van Niel's course provided a superb historical review of microbiology and a powerful antidote to medically oriented bacteriology. He dwelled on the good microbes in the environment and forbade the mention of pathogens, except those few that figured prominently in the history of microbiology. Progress was described in the exploits of his heroes: Anton van Leeuwenhoek, Louis Pasteur, Sergei N. Winogradsky (1856–1953), the Russian soil microbiologist; and not least, the yeast cell. Van Niel traced his own Dutch lineage with reverence to Albert J. Kluyver (1888–1956), his teacher, and farther back to Martinus W. Beijerinck (1851–1931), who taught Kluyver. Van Niel's own work on bacteria had provided a major insight into photosynthesis. He showed that the splitting of a water molecule by light energy generates oxygen and releases energy, which can be stored and used to convert carbon dioxide to sugar later in the dark.

Once during the course I gave a seminar on my previous work on the isolation of enzymes from the cellular juices of microbes. Afterwards, van Niel told me: "This is beautiful work. I know it needs to be done. I myself would not have the heart to grind up the little beasties."

Spontaneous Generation

One of the stories I told my boys was about Anton van Leeuwenhoek (1632–1725), the grandfather of microbiology. Income from his dry-goods store in Amsterdam and some minor municipal jobs allowed him ample spare time to grind and polish very small biconvex lenses less than one-eighth inch wide. Mounted between metal plates, these lenses made simple but very effective microscopes that magnified 200-fold (Fig. 4-12). He made over four hundred such magnifying glasses. He was inventive, curious, industrious, and persistent. He examined everything in reach.

In a droplet of old rainwater he could see many tiny "animalcules" moving about. A richer source was his decayed tooth. As he described it:

> I took this stuff out of the hollows in the roots and mixed it with clear rainwater, and set it before the magnifying glass to see if there were as many living creatures in it as I had aforetime discovered in such material: and I must confess that the whole stuff seemed to me to be alive. But notwithstanding, the number of these animalcules was so extraordinarily great (though they were so little withal, that 'twood take a thousand million of some of 'em to make up the bulk of a coarse sand grain, and though several thousands were a-swimming in a quantity of water that was no bigger than a coarse sand grain is), yet their number appeared even greater than it really was because the animalcules, with their strong swimming through the water, put many little particles which had no life in them into a like motion, so that many people might well have taken these particles for living creatures too.

The "little beasties," as he called the bacteria, had regular shapes: spheres, rods, and spirals (Fig. 4-13). They were easily distinguished

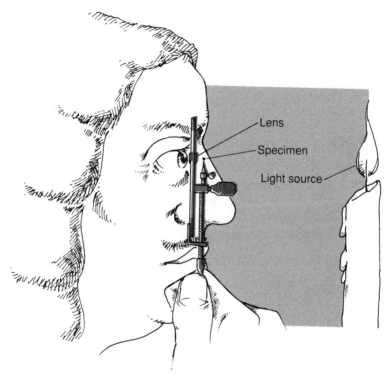

Lens

Specimen

Light source

FIGURE 4-12
Van Leeuwenhoek looking at a specimen magnified about 200-fold by the lens of one of his microscopes, with illumination provided by a candle. Magnification near 1,000-fold is obtained with present-day light microscopes.

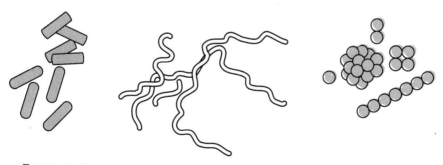

FIGURE 4-13
The great variety of bacterial shapes include rods (bacilli), spirillar forms (spirochetes), and spheres (cocci).

from inanimate dust and dirt particles. In the course of fifty years of exceedingly careful observations communicated to The Royal Society of London, van Leeuwenhoek was also the first to see the round red blood corpuscles in human blood and the oval cells in the capillaries of frogs and fish. He was the first to observe spermatozoa, protozoa, and vinegar eelworms and to reconstruct the complex life cycle of an ant, from egg to larva to pupa to adult. He realized that the smallest living creatures must emerge from similar living creatures rather than from decaying matter, as was generally thought at the time. The convincing proof that even the tiny bacteria were not generated spontaneously had to await the experiments of Lazzaro Spallanzani and Louis Pasteur in succeeding centuries.

It is hard to imagine now, but in van Leeuwenhoek's time people still believed that beetles and wasps were generated out of cow dung and swarms of mice were begotten by the mud of the Nile. For many centuries, Aristotle and his disciples had taught that some animals were born from soil, plants, and mythical creatures and that maggots were generated by putrefying meat. One questioning mind belonged to Francesco Redi (1626–1697), poet, lexicographer, and court physician to the grand dukes of Tuscany. I described to my boys the simple experiment he had performed. Redi covered a container of putrefying meat with fine Naples gauze. Despite exposure to air and warmth, the spoiled meat inside the container never developed maggots. Instead, Redi saw them come out of eggs laid by flies on top of the gauze.

While the spontaneous generation of mice and flies could be dismissed, the source of the primitive microbes was another matter. In 1749 John Needham, a Welsh naturalist, concluded that mutton gravy possessed a "vegetative force" to generate microbes. He boiled the broth and destroyed the microbes; yet they reappeared in great numbers, even in stoppered flasks. But Lazzaro Spallanzani (1729–1799), an Italian abbot and naturalist, was not convinced. He realized that the cork stoppers Needham used to close his flasks might not have been airtight and thus may have allowed microbes to enter with the air when the heated vessels were cooled. When he melted and sealed the necks of his glass flasks, the heated meat broths remained sterile indefinitely. He then did a striking experiment. From a broth with rod-shaped microbes, he took a droplet and managed under the microscope to tease one of the rods into an empty droplet

FIGURE 4-14
A simple vessel for bacterial growth is a flask
or test tube kept sterile by a cotton plug.

and then patiently waited and watched for an hour; he saw the rod
lengthen and divide in two.

In time, though, other scientists raised objections to Spallanzani's
experiments. They claimed that the lack of air in his sealed flasks
prevented the broth from generating microbes. These objections
were in turn answered in several ways, most simply by admitting air
to a flask through a cotton plug (Fig. 4-14). With this device, used
routinely to this day for filtering out minute particles, no microbes
appeared in heated solutions.

Spallanzani's work should have settled the issue of spontaneous
generation, but some ideas die hard. After several decades, this be-
lief began to flourish again, and it took the combined polemical and
experimental skills of Louis Pasteur in the mid-nineteenth century
to subdue it. To show that yeast cells are indispensable for fermentation, he devised a flask with a long, narrow, gooseneck (Fig. 4-15).
The "natural" air could enter this flask freely, but microbes and dust
settled in the gooseneck and did not contaminate the solution. The
broth remained free of fermentation unless he splashed a bit of it into
the neck and carried back some of the microbes and dust. When he
opened flasks in the dust-free air of a deep, dank cellar or in the
crystalline air of Mont Blanc, no microbes populated the broth.

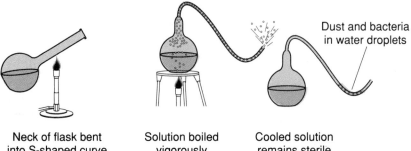

Neck of flask bent   Solution boiled   Cooled solution
into S-shaped curve   vigorously   remains sterile

FIGURE 4-15
One of Pasteur's classic experiments that helped disprove spontaneous
generation of microbes depended on maintaining the sterility of a boiled
nutrient medium by using a gooseneck flask to trap the dust and bacteria
that would otherwise be drawn in from the air upon cooling.

As Pasteur, in 1864, described the sterile broths:

> For I have kept from them, and am still keeping from them, that one
> thing which is above the power of man to make; I have kept from them
> the germs that float in the air. I have kept from them life . . . There is no
> condition known today in which you can affirm that microscopic be-
> ings come into the world without germs, without parents like them-
> selves. They who allege it have been the sport of illusions, of illmade
> experiments, vitiated by errors which they have not been able to per-
> ceive and have not known how to avoid.

Germs Cause Disease in Wine and People

Louis Pasteur's extraordinary exploits, as described in René Dubos's
glowing biography of 1950, were a wonderful source of germ stories.
Pasteur's contributions and those of Robert Koch, a younger contemporary, have meant so much to biologic science and are so fundamental to everyone's understanding of life processes that I must
pause in this narrative to savor them.

Pasteur's origins were humble. He grew up in Arbois, a small town
in the east of France, where his father was a tanner. An industrious
but undistinguished student, he made his way to Paris, became fascinated by chemistry, and solved one of the great problems of
chemistry and physics when he was twenty-six.

Pasteur was struck by the observation of a German chemist, Eil-hardt Mitscherlich, regarding two forms of tartaric acid (tartaric and paratartaric) obtained from the tartar in wine fermentation vats. They were identical in chemical composition and yet were different in that tartaric acid rotated a beam of polarized light to the right, whereas paratartaric had no effect at all. How could identical sub-stances behave so differently? Upon examining the crystals of the two forms, he noticed things that the many chemists before him had missed. There were facets on the crystals, and they were not sym-metrically arranged. The crystals of the tartaric acid form all looked alike, but in paratartaric acid there were two kinds—one identical to those of tartaric acid and the other its mirror image. When he had sorted out the two kinds of paratartaric acid crystals, there were equal numbers of them; those identical to tartaric acid rotated polarized light to the right, whereas their mirror images rotated the light to the left. Now he could explain why with paratartaric acid, in which equal numbers of the right- and left-rotating forms neutralized each other, there was no net rotation of polarized light. Pasteur rea-soned correctly that the crystal forms reflected a comparable mirror-imagery in the structure of the molecules that made up the crystal. This single discovery brought him instant fame and is recognized today as one of the great insights into chemical structure.

Pasteur became the father of microbiology and the major force in developing the germ theory of disease because his training as a chemist led him to apply the rigor of physical methods to his biolog-ical experiments. In view of the wide separation between the disci-plines of microbiology and chemistry for most of the twentieth cen-tury, it is worth recalling that this foremost microbiologist was described in the 1911 edition of the *Encyclopaedia Britannica* as a chemist and the acknowledged head of the greatest *chemical* move-ment of his time.

In Lille, where Pasteur had been made professor and dean of the Faculty of Sciences, he was asked to help with a serious problem in the sugar beet distilleries. Instead of alcohol, the fermentation vats were often filled with a slimy fluid that smelled and tasted like sour milk. Pasteur examined samples from the good and bad fermenta-tions under the microscope. Almost without fail, the healthy fer-mentations yielding alcohol had large numbers of growing and di-viding globular yeast cells; those that had gone sour lacked them and instead accumulated gray deposits swarming with short, rod-shaped particles. He was able to cultivate these rods in a clear, nutrient bouillon, less complex than sugar beet juice, and concluded that these tiny creatures made sugar into the lactic acid of sour milk, unlike the yeasts, which converted it to alcohol.

There were occasions when lactic acid was not produced; in its place was butyric acid, which exuded the unpleasant odor of rancid butter. In these cases, he saw large rods rapidly moving about, rather than the small quivering ones. In opposition to the prevailing views of the leading chemists, such as Jöns Jakob Berzelius of Sweden and Justus von Liebig of Germany, Pasteur held that these living microbes were the factories in which all these fermentations took place, rather than the by-products of chemical processes outside them. With his gift for picturesque language, he explained that the millions of gallons of wine in France and the oceans of beer in Germany were not made by men but rather by the incessant labor of "armies of creatures ten-billion times smaller than a wee baby."

Pasteur's well-publicized exploits in science and industry now had wide recognition, and he was called upon to diagnose and cure diseases that were devastating wine and vinegar fermentations. To his native Arbois, he took his laboratory equipment and staff to investigate why so many wines were turning bitter, ropy, or oily. In each instance, he could identify successful wine fermentations with the presence of yeast and the failures with the intrusion of other distinctive microbial populations.

In the vats producing vinegar commercially, a thin surface film is responsible for combining the oxygen from the air with the alcohol of wine to form acetic acid. Pasteur discovered that when the microbes that made up the surface film became submerged, they lacked sufficient oxygen to convert the alcohol; with too much oxygen, they destroyed the formed acetic acid by oxidizing it further to carbon dioxide and water.

Lacking a deeper understanding of how the invasion by unwanted microbes spoiled a fermentation, Pasteur was still able to devise an effective means of preventing these aberrations. Gentle heating of the wine (to about 145°F for half an hour), just after the fermentation was finished, killed most troublesome microbes but did not affect the flavors. This process of partial sterilization was widely adopted for the preservation of wine, beer, vinegar, cider, milk, and many other perishable foods and beverages and has become known as pasteurization.

Showing germs to be the responsible agents in deviant fermentations did not immediately implicate them as the cause of human disease. Throughout the last century, puerperal (childbed) fever mysteriously killed many women after childbirth. Might the disease be caused by a germ carried by midwives and physicians from one mother to another? Among the few who believed this was Oliver Wendell Holmes (1809–1894), a physician, literary figure, and father of the famous jurist. Sharing the same conviction was the Hungarian physician Ignaz Philipp Semmelweis (1818–1865), who sharply reduced the incidence of infection in his obstetrical patients by observing better hygiene and using antiseptics. Nevertheless, the prevailing view in medical circles was that puerperal fever, typhoid fever, tuberculosis, and other diseases were not contagious. While Pasteur's germ theory might apply to wines and beer, physicians did not believe that germs were responsible for illness in people.

Even when microbes were found in diseased tissues, their presence was regarded as a secondary consequence of the weakened state of the tissue, possibly an aggravating element but not the primary cause of illness. Florence Nightingale, who is justly celebrated for pioneering in sanitation, hospital practices, and nursing care and most of all for statistical rigor in collecting data, expressed the common skepticism regarding germs as the agents of disease. Based on extensive experience in military hospitals during the wars in Crimea and India, and obsessed with the importance of hygiene, she wrote:

> I was brought up by scientific men and ignorant women distinctly to believe that smallpox was a thing of which there was once a specimen in the world, which went on propagating itself in a perpetual chain of descent, just as much as that there was a first dog (or first pair of dogs) and that smallpox would not begin itself any more than a new dog would without there having been a parent dog. Since then I have seen with my eyes and smelled with my nose smallpox growing up in first specimens, either in close rooms or in overcrowded wards, where it could not by any possibility have been 'caught' but must have begun. Nay, more, I have seen diseases begin, grow up and pass into one another. Now dogs do not pass into cats. I have seen, for instance, with a little overcrowding, continued fever grow up, and with a little more, typhoid fever, and with a little more, typhus, and all in the same ward or hut. For diseases, as all experiences show, are adjectives, not noun substantives . . .
>
> The specific disease doctrine is the grand refuge of weak, uncultured, unstable minds, such as now rule in the medical profession. There are no specific diseases: there are specific disease conditions.

Koch's Postulates: One Germ, One Disease

It took the genius of Robert Koch (1843–1910), an eccentric German country doctor, to establish that it takes a particular germ to cause a particular human disease. To provide some diversion from his frustrating general practice, Koch's wife gave him a microscope on his twenty-eighth birthday. Divert him it did. Finding, examining, and propagating microbes became a consuming passion and made him the greatest pure bacteriologist of all time. He discovered the large rod that causes anthrax, the tiny one responsible for tuberculosis, and the comma-shaped rod ("vibrio") of cholera.

Although the history of an idea can usually be traced to antiquity, an idea matters most when its value has been emblazoned by some recent or current application. Such was Koch's demonstration, only a hundred years ago, that a complex phenomenon, the disease anthrax, can be rigorously proven to be due to a single root cause, one particular microbe. The demonstration of so clear a causal relationship in something as dramatic as a human disease has by extension had an impact on other phenomena in science and social affairs.

Anthrax (gangrene of the spleen) is a ghastly disease that kills sheep, goats, and cows, singly or in herds, and has often claimed the lives of the farmers, sheepherders, wool gatherers, and hide dealers who tended them. In the blood and spleen of animals and people dead of anthrax, Koch invariably found distinctive rod-shaped bacteria, sometimes connected in long chains or threads. Others had made similar observations but had only conjectured about a causal association between rod and disease.

Koch went much further. The mice he injected with infected tissue promptly developed anthrax and had these rods in their blood and spleen. Tissues of normal animals or people neither had these microbes nor caused the disease when injected into mice. Were these microbes alive and really the cause of anthrax? Koch suspended a tiny bit of infected tissue in a droplet of nutrient medium made up of clear fluid from ox eyes. Within hours, he was astonished to see the rods multiply a million-fold. With a few rods taken from the droplet, he seeded a fresh sterile droplet and once again observed their enormous multiplication. Even after eight such transfers, the culture fluid teamed with rods, and a little sample was sufficient to produce anthrax in healthy mice or sheep. The rods could now be classified in the genus *Bacillus* (Latin *baculum*, rod), with the species designation *anthracis* attached.

By these experiments with anthrax, repeated over and over with many variations, Koch developed the protocol required for the rigorous proof that a given microbe causes a particular disease:

(1) Find the microbe invariably where one finds the disease.

(2) Cultivate the microbe in large numbers (for example, a million-fold increase) as a homogeneous population in a nutrient medium.

(3) Use the culture to produce the disease in a susceptible animal.

(4) Recover the distinctive microbe again, in pure culture, from the experimentally infected animal.

These criteria, known since as Koch's Postulates, were difficult to satisfy because the cultures of rods in a nutrient broth were generally mixed with other microbes and on occasion even outgrown by them. How to obtain a microbe in pure culture?

By adjusting the nutrients, the acidity, temperature, and degree of aeration, Pasteur had learned how to favor the growth of one microbe over another and thus permit it to populate a culture or fermentation. However, even one microbe persisting among a billion others could under altered circumstances outgrow and dominate the rest. The arithmetic consequences of microbes doubling in number every twenty minutes are staggering. In ten hours, one microbe can generate a billion, weighing altogether about one milligram (one thousandth of a gram). In twenty-four hours, if the supply of nutrients could be maintained and all wastes removed, the microbial mass would reach ten thousand tons. Given a few more hours under these ideal circumstances, it would exceed the mass of the earth.

The solution in 1881 to the problem of obtaining absolutely pure cultures—that is, populations derived from a single organism—proved to be astonishingly simple. The technique became more indispensable than the most powerful microscope and is still used today in every research, industrial, and hospital laboratory. It was discovered by accident, but Koch's mind was prepared to exploit it.

One day Koch glanced at half of a boiled potato and saw several specks on its flat surface. It was the variety of their colors and shapes that attracted his attention. In the habit of examining everything under his microscope, he was startled to find that a gray-colored speck had only rod-shaped microbes, a yellow speck was made up exclusively of spherical bodies, and a tiny red mound yielded only wriggling little corkscrews. Koch realized at once that these specks were pure cultures, each derived from a single germ floating about in the air and landing by chance on a solid nutrient, in this instance a potato. To produce a solid nutrient with a more uniform surface, he mixed gelatin into beef broth and poured a layer on a glass plate to set. A sufficiently diluted mixed culture of microbes spread over the solid surface produced discrete and distinctive colonies made up of a billion identical progeny (clones) from a single microbe.

Aside from two further refinements, the use of solid nutrient media for isolation of pure microbial cultures has undergone no substantial change in the past hundred years. One improvement was finding a better solid medium than gelatin, which liquefies near body temperature and so cannot be used in a warm climate or for microbes that require warm temperatures. Frau Hesse, the wife of one of Koch's coworkers, had used agar-agar as a gelatin substitute in making jams. Dutch friends brought her some from Batavia, where it is derived from algae in the east Asiatic seas. Agar-agar (or simply agar) has the fortunate property of requiring a temperature near boiling to melt it, but once melted it can be cooled to near 40°C before it sets into a stiff and nearly transparent solid.

The other improvement, made by one R. J. Petri, an assistant of Koch, was replacing the flat glass plate with a shallow, round, covered dish (Fig. 4-16). In these Petri plates (or dishes), a culture could be examined repeatedly with minimal contamination; stacks of them could be stored conveniently. The agar Petri plate opened the gates for a rush of bacteriologists to discover the microbes responsible for

FIGURE 4-16
The Petri dish contains a solid nutrient medium upon whose surface individual bacteria can multiply to form colonies (or clones). These transparent glass (or plastic) dishes are easily sterilized, stacked, and examined.

every disease. The nutrient base of the agar medium was manipulated endlessly by addition of various sugars, proteins, vitamins, minerals, blood, and tissue extracts to favor or detect the growth of particular microbes. The technique made it possible to measure the abundance of each microbe in water, air, soil, and body fluids.

Consumption, Cholera, and Other Conquests of Microbe Hunting

Of the forty-four students who started medical school in my class at the University of Rochester in 1937, five contracted tuberculosis in the first two years, an incidence similar to that of previous classes. They stayed in sanitoria for bed rest and fresh air for a year or more; effective antibiotics were still ten years away. Most of the medical students recovered; some did not. Why this appalling contagion?

Fifty years earlier, one of every seven deaths in Europe and the United States was from consumption, as tuberculosis was then known. In 1882 this Great White Plague became the next microbial safari for Robert Koch after his success with the bacillus of anthrax. He did not doubt the contagiousness of tuberculosis or its microbial origins, inasmuch as the disease was transmissible to experimental animals. But he failed repeatedly in trying to find any organisms in the tiny, grayish-yellow, cheesy nodules (called tubercles) in affected tissues of diseased patients.

Koch was aware that various dyes can make microbes more visible and distinguishable. But staining did not make the microbe of tuberculosis visible until he discovered that microbes spread on a glass slide take up dyes more readily after they are dried and heated ("fixed"). In tubercles treated this way and then exposed in alkaline solution to methylene blue dye, he saw for the first time masses of very tiny, wispy, bent blue rods. Thereafter, he could always find these bacilli in tuberculous tissues, in the coughed-up sputum of patients with affected lungs, and in a variety of infected experimental animals, but never in normal people or animals.

Despite the invariable association of these bacilli with tuberculous disease, Koch could not regard them as a certain cause of the disease, because he had been unable, using all known culture media, to propagate infectious bacilli outside the animal body. Once again

Koch's ingenuity and persistence prevailed. He devised a medium that would resemble the animal interior. A jelly of blood serum, clotted and solidified in a test tube set at an angle, provided a large, slanted surface on which the tuberculous material was streaked. He waited many days, long after anthrax and other bacilli would have achieved luxuriant growth. Only after two weeks did colonies of the tubercle bacilli begin to appear. With these cultured bacilli he produced tuberculosis in experimental animals, and from them he was able to culture the virulent bacilli that would later be called *Mycobacterium tuberculosis.*

How do people contract the disease? Plausibly, Koch thought, the microbes are inhaled in the spray from coughs and sneezes of diseased patients. To test this hypothesis, Koch demonstrated that guinea pigs and rabbits that were repeatedly exposed to aerosols (mists) of cultures of the tubercle bacillus came down with the disease. The danger of pumping mists of deadly microbes into boxes and handling the animals and apparatus was enormous. Had Koch lived to ninety, he could have had this dangerous experiment done for him in another context—by medical students.

In the pathology course during our second year at Rochester, we medical students assisted in the autopsies of tuberculosis cases at the nearby Monroe County Hospital. We were urged to examine closely—practically to sniff the spray from—lungs freshly sliced to expose the tubercles; as a result many of us contracted tuberculosis. How strange and inexplicable that Dr. George H. Whipple, head of the pathology department and dean of the University of Rochester Medical School, famous as a Nobel laureate and revered by many for his medical wisdom, could have been a witness to, and perhaps even responsible for, this hazardous and avoidable exposure. The bravado and hazing in physician training, along with public indifference toward occupational safety at that time, may have been in part responsible.

Cholera has vied with tuberculosis as a major scourge over the centuries. Imagine this scenario in Alexandria, Egypt, in 1883: The dreaded Asiatic cholera, which had swept into Europe in several previous pandemics, killing hundreds of thousands in its path, is raging through Egypt en route from its usual nesting place in India. Onto the scene come the world's two foremost teams of microbe hunters to compete for the discovery of the microbe responsible for cholera.

Beyond the rivalry for the prize of this major scientific achievement are pent-up personal hostilities and patriotic loyalties in the contest between the French and German commissions. Dying patients, corpses, menageries of experimental animals, microscopes, and the steaming heat form a backdrop; then a sudden recession of the epidemic alters the tempo. The gifted young French scientist Louis F. Thuillier dies of a fulminating cholera infection, and both groups return home disconsolate, with inconclusive results.

A fulminating cholera epidemic in Calcutta revives the indomitable Koch. He goes there and finds a comma-shaped rod in every one of forty people dead of cholera. He finds it in linen soiled by their rice-watery stools, in the intestines of patients stricken with the disease, and in water tanks from which they drank. He obtains pure cultures of the infectious *Vibrio cholerae* on a beef broth medium and sails back to Germany for a hero's welcome.

With the identification of the microbial villains for each of these diseases came the means to avoid them by improved hygiene and vaccines and, later, by treatments with specific serums and antibiotics. Improved techniques developed in the discovery of the anthrax, tuberculosis, and cholera bacilli were successfully applied to diphtheria, scarlet fever, and typhoid fever. It became open season for microbe hunters until it seemed, by the 1970s, that every pathogenic microbe that had ever evolved was by then known and identified. Yet in the past decade, three illnesses have appeared which produced high fatality rates and, at first, seemed mysterious in their origins and mode of dissemination. For each of the three—Legionnaire's disease and toxic shock syndrome, both of bacterial origin, and acquired immune deficiency syndrome (AIDS), caused by a virus—the responsible agent has been tracked down, though in the case of AIDS no cure for this devastating disease, nor vaccine to prevent it, has yet been found.

In some instances, the early microbe hunters were thwarted by initial failures which led eventually to major discoveries about microbial disease. It was found, for example, that some microbes, such as those causing anthrax, tetanus, and botulism, can anticipate adverse conditions by becoming spores with a capacity to hibernate for years and withstand lethal extremes of heat and radiation. A second major discovery among disease-producing agents was the viruses, one-thousandth the size of a bacterium, much too small to be seen in the light microscope. In the six years that I was chairman of the

Department of Microbiology at Washington University and officially a microbiologist, what fascinated me most were the heat-resistant spores and the tiny viruses. These phenomenal objects were to become focal elements of my biochemical research during the next thirty years.

Chapter 5

The Synthesis of DNA

In 1953 James Watson and Francis Crick reported their startling discovery that DNA is a pair of chains spiraling about each other—a double helix. This structure not only could account for the physical properties of DNA but also could explain how DNA might be faithfully reproduced in a cell. Measured in significance per word, their very brief paper in *Nature* is surely one of the most important reports ever published in biology. Yet no gem is flawless. In their paper, Watson and Crick naively proposed that replication of DNA would proceed *spontaneously* were the building blocks to align themselves along the parental chains serving as templates. There is little likelihood that any cellular reaction, let alone one so intricate and vital as the rapid and faithful replication of DNA, could proceed without the catalysis, direction, and refined regulation that enzymes provide.

Within two years of Watson and Crick's historic report, I had found, in juices extracted from cells, an enzyme that synthesizes the huge chains of DNA from simple building blocks. On the basis of this chronology, it is commonly (and understandably) assumed that the Watson–Crick discovery spurred me to search for the enzymes of replication. But that is not the way it happened. In 1953 DNA was far from the center of my interests. The significance of the double helix did not intrude into my work until 1956, after the enzyme that assembles the nucleotide building blocks into a DNA chain was already in hand. My interest in the replication of DNA, the focus of my research for the past thirty years, developed primarily from my unremitting fascination with enzymes.

Whence Orotic Acid?

My ambitions soared in 1950. I had earlier found an enzyme in extracts from potatoes that inserts a water molecule into a respiratory coenzyme, cleaving it into two nucleotide components. By using this apparently ordinary enzyme to produce a novel starting material, I could discover how coenzymes are made in cells and with it a basic, biochemical theme used in the biosynthesis of lipids, steroids, proteins, and nucleic acids. Having learned how a nucleotide, the building block of RNA and DNA, is built into a coenzyme, I was led to wonder if many thousands of them might be assembled to make the chains of the nucleic acids. First, I had to know how the nucleotides themselves were made from still simpler molecules. With Buchanan and Greenberg going after the purine nucleotides (A and G), I would pursue the biosynthesis of the pyrimidines (U, C, and T).

Tracing the fate of a molecule by labeling with a radioactive isotope of phosphorus or carbon, then a relatively new technique, was indispensable to finding the multiple routes of nucleotide biosynthesis. Separating nucleotides and the enzymes that make them by newly developed chromatographic methods was equally important. It is difficult for me now to recall, and impossible for my students to believe, that anything could have been discovered without tracer and chromatographic techniques. Their application has made it possible for us to delineate the intricate and awesome pathways which molecules travel on their way to becoming part of RNA and DNA. And with this basic knowledge have come practical benefits: the improved diagnosis and treatment of many diseases, including gout and cancer, and control of the body's rejection of organ transplants.

Around 1950 biochemists still believed that the metabolic traffic of breakdown and synthesis ran on the same track. The pathway of degrading sugar to alcohol (in yeast) and lactic acid (in muscle) was thought to be largely used in the reverse direction to make sugar from these simpler substances. Inasmuch as each of the dozen steps in the pathway was catalyzed by an enzyme and virtually all of the steps appeared to be reversible, biosynthesis of a carbohydrate was expected to be simply the reverse of its breakdown. By extension, the same should be true for the biosynthesis of proteins and nucleic acids. This preconception caused me to look for bacteria in soil proficient in using a pyrimidine as a fuel for energy and as a source of carbon and nitrogen for growth. I hoped their facility in breaking down a pyrimidine might reveal key steps in its biosynthesis.

Another reason for resorting to enrichment cultures at that time was the arrival, as I have mentioned, of Osamu Hayaishi, who had considerable experience with enrichment cultures before he came to work with me as a postdoctoral fellow. Born in Stockton, California, when his father was in clinical training, he returned to Japan at the age of two. After serving as a physician in the Japanese navy during the war, he sought training in biochemistry. His first postwar stop in the United States was at the University of Wisconsin in Madison, where he worked on aerobic phosphorylation in the laboratory of David E. Green. There he became Sam Schnellstein to his Jewish friends—Sam for Osamu, Schnell for *haya* (quick), and stein for *ishi* (stone). Hayaishi later moved with me from the National Institutes of Health to Washington University to become an assistant professor, the first appointment I made as the chairman of the new department of microbiology. He returned briefly to NIH before accepting the prestigious professorship of biochemistry in Kyoto and becoming in subsequent years the foremost biochemist in Japan.

When Hayaishi and I isolated pure cultures of soil bacteria that avidly consumed pyrimidines and thrived on them as their sole nutrient, we were startled and dismayed that in their metabolism pyrimidines were oxidized to barbiturates, more familiar as sedative drugs than what we expected as reasonable precursors in pyrimidine biosynthesis. It seemed that, in this instance, the degradative pathway of pyrimidines was no reflection of the route to their synthesis.

I turned to orotic acid as another way of attacking the problem of pyrimidine biosynthesis. Orotic acid is uracil with a carboxyl (CO_2) group attached (Fig. 5-1). Named after whey (Greek *oros*), in which it was discovered by two Italian chemists at the turn of the century, orotic acid was forgotten for almost fifty years. It turned up again as a massive accumulation in culture media of a mutated bread mold

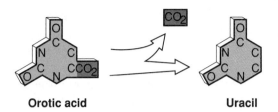

Orotic acid   **Uracil**

FIGURE 5-1
Upon the loss of a molecule of carbon dioxide, orotic acid
would become uracil, one of the four bases of RNA.

(Neurospora). Still later, it was found in the urine of infants with a
fatal inborn metabolic error that proved to be similar to that in
Neurospora.

Presumably, the mutants (both *Neurospora* and human) were
deficient in the enzyme that converts orotic acid to the next intermediate in a metabolic chain. That enzyme might possibly be one
that converted orotic acid to uracil by removing the carboxyl group,
in which case orotic acid might be an important intermediate in the
synthesis of uracil. This supposition was strengthened by experiments which tracked the fate of isotopically tagged orotic acid fed to
normal *Neurospora* or rats or taken up by slices of rat liver. The
labeled orotic acid was converted to the pyrimidines of RNA and
DNA.

Still, the progress of orotic acid on a metabolic pathway to nucleic
acid pyrimidines does not prove it to be a precursor on the main
track of biosynthesis; orotic acid might be connected to the track by a
spur used only when the substance is supplied to the cell from the
medium around it. To settle this question, I tried to discover the
mechanisms and magnitude of the manufacture of orotic acid within
the cell itself. Once again, I looked for clues in the metabolic breakdown of orotic acid. When I failed to find any such activity in extracts of various animal tissues, I arranged to visit H. A. Barker's
laboratory in Berkeley, where with his guidance I isolated a bacterium genetically equipped to use orotic acid for its energy and
growth needs (see Chapter 4). I returned with this prize catch to NIH
to try to extract cell-free preparations and then purify the enzymes
which would carry out the metabolism of orotic acid.

Soon thereafter, Irving Lieberman joined my laboratory. Lieb, in the years I knew him, came closer than anyone I have worked with to being a "mad scientist." Nothing daunted or deterred him. His energy was unflagging, his output prodigious. Unhappy with his doctorate in veterinary medicine, he was completing his PhD in microbiology under Barker when I met him. On the day Lieb arrived at my laboratory in Bethesda to start his postdoctoral fellowship, it was already near quitting time. He had driven across the country with his wife and infant son and come directly to the laboratory. With barely a greeting, he wanted to know about research projects he might work on, and he instantly chose orotic acid. I guessed it would take a week or so to find housing and get settled and offered to help. Lieb interrupted, "Can we go over an experimental plan now?" I was already late for dinner at home. "Sure," I said, "the first thing tomorrow." "Why tomorrow, Doc? We can get some cultures started tonight." I could not dissuade him from sterilizing media and starting a series of culture flasks right then and there. Later I learned that his wife and child had been waiting for him in their baggage-stuffed car those several hours in a darkening, abandoned parking lot.

We failed many times in extracting from the disrupted bacteria a juice that was active in disposing of orotic acid. How foolish I felt when it dawned on me that the very first step in the process might reflect the anaerobic lifestyle of this bacterium. When we fortified the juice with the NAD coenzyme in the reduced form, as opposed to the more familiar oxidized form predominating in aerobic bacteria, the removal of orotic acid proceeded vigorously. Upon separating the enzymes responsible for each step in the disassembly of orotic acid, we found that carbamyl aspartic acid, a derivative of the amino acid aspartic acid, was produced (Fig. 5-2). This was a plausible intermediate in a biosynthetic pathway to orotic acid.

At this juncture, I felt that orotic acid was likely to be a genuine precursor in the cellular manufacture of the pyrimidine building blocks of RNA and DNA. We were at a crossroads. We could try to demonstrate the synthesis of orotic acid from aspartic acid via carbamyl aspartic acid in extracts of yeast cells and liver. On the other hand, I was also eager to find how orotic acid was converted to a genuine pyrimidine building block of nucleic acids. Lieberman too was pushing for us to go for "orotic up" (assembly into nucleic acids) rather than "orotic down" (to its origins from smaller molecules) (Fig. 5-3). So we took the "orotic up" road.

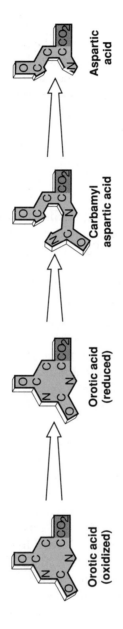

FIGURE 5-2

Steps in the metabolic breakdown of orotic acid. Orotic acid, modified by addition of two hydrogens (that is, reduced), is split by a molecule of water to form carbamyl aspartic acid, which is then converted to an amino acid (aspartic acid) by the removal of the carbamyl (N-C-O) group.

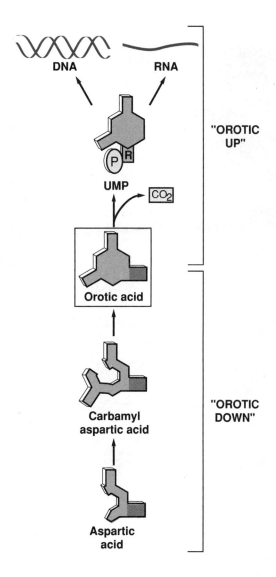

FIGURE 5-3
The metabolism of orotic acid can be examined
either by looking "down" to its origins (aspartic
acid) or "up" to its conversion to UMP, a building
block of RNA and a precursor of DNA.

FIGURE 5-4

The Enzyme Section at the National Institutes of Health in 1952. *Standing:* O. Hayaishi, R. Stroud, H. A. Barker, I. Clark, B. Verham, B. L. Horecker, I. Lieberman, R. Clary, R. Hilmoe, A. Kornberg, H. Klenow. *Kneeling:* L. A. Heppel, P. A. Smyrniotis, W. E. Pricer.

Leaving NIH, My Alma Mater

These orotic acid experiments were not the only crossroads I faced in 1953. I had to decide whether to continue at NIH or to veer westward to an academic life at Washington University in St. Louis, a city that would later become my "gateway to the (far) west." The decision to go west was based on two considerations, both of which turned out to be errors in judgment. First, I believed that the advent of the Clinical Center and the disease-oriented institutes would stifle basic research at NIH. And second, I believed that administrative life in a university would be more inspiring than life at NIH. As it turned out, research at NIH flourished, and I learned to my dismay that university administration and politics can indeed be burdensome.

When I came to NIH in the fall of 1942, Thomas Parran, who was then Surgeon General, told me that he wanted his Public Health Service officers less well-rounded but sharper at the edges. So an assignment to NIH promised to be more than the usual two-year military service rotation. While I was still in uniform, NIH sponsored my training in biochemistry with Severo Ochoa in New York in 1946 and then with Carl and Gerty Cori in St. Louis. When I returned in 1947, the Institute had not changed much since the prewar years. But the next year an *s* was added after *Institute* on the lettering of the Administrative Building architrave, signifying the creation of dental, heart, and other categorical institutes. The Research Grants Division was established, and planning for the Clinical Center was started.

Despite these harbingers of change, the five years between 1947 and 1952 were the most productive and gratifying in my scientific life. I recall a visit in 1950 from Gerty Cori. She lamented my being in a government laboratory. How could I persuade her that, in working without distraction on problems of my own choosing and digesting biochemistry publications every noon hour with my close friends and colleagues Bernie Horecker, Leon Heppel, and Herb Tabor (whose background and outlook were so like mine), I was enjoying an ideal academic environment (Fig. 5-4)?

Few people recognize the NIH acronym, nor do they know of the National Institutes of Health when it is spelled out for them. Yet the achievements of NIH defy exaggeration. It is the prime source of the most extraordinary revolution in biologic science. As expressed by Lewis Thomas: "All by itself, this magnificent institution stands as the most brilliant social invention of the twentieth century, anywhere." More than any university, NIH is my alma mater, and so I feel impelled to interrupt the narrative of this memoir to relate its origins, how it came—to quote Thomas again—"to do something unique, imaginative, useful and altogether right," and to explain why it has been and remains vital for the future of medical science.

In 1987 we celebrated the centenaries of the founding of both the National Institutes of Health and the Pasteur Institute of Paris. Unlike the impressive building, staff, and worldwide recognition the Parisian institute enjoyed at its inception, NIH's beginning was humble in the extreme, a one-scientist, one-room laboratory in the attic of a Public Health Service Marine Hospital on Staten Island. From that start, NIH grew at a modest rate, occupying six small buildings in Bethesda when I arrived there in 1942. Then came the explosive expansion in the post-World War II decades which changed the face of medical science—13,000 people working in fifty buildings on a 300-acre site in Bethesda, along with 52,000 scientists at 1,600 institutions around the world, supported by an annual budget of near $7 billion.

The colossal achievement for which NIH is justly famous is the innovation and maintenance of this vast program of grants for support of research and training in laboratories throughout the world. This program has been the single most important foundation for the biological revolution of the postwar period. Guided initially by NIH scientists, the peer-review system for awarding grants and fellowships has administered many tens of billions of dollars with a scrupulous regard for quality and without a hint of chicanery. I know of no government program of this magnitude with such a magnificent record.

Sometimes overlooked is an achievement of NIH that prompted a large group of us to assemble in Bethesda in 1975 for the first alumni reunion. The talk I gave on that occasion tried to express that NIH, more than any college or university, had shaped our lives in the most profound way. In the untrammeled atmosphere of well-equipped, well-managed laboratories, young MDs and PhDs were introduced to professional science. Some remained at NIH, but well over 25,000 left to staff research, clinical, and administrative departments throughout the world. Today they populate and—as professors, chairmen, and deans—direct the finest university departments of basic medical science and clinical science. They are the clinicians in the leading hospitals, the research directors of the foremost pharmaceutical companies. They bring a novel outlook from their training in basic biological and chemical sciences to the lecture hall, laboratory, bedside, and factory. NIH is truly a National *University* of Health.

Looking back at this remarkable success, I am impressed, as an early participant, that this astonishing institution began its development without a plan or the leadership of any one individual. If a proper historical account is ever written, several names and some key policy decisions will be cited. Among the most significant people mentioned will be James A. Shannon, who directed NIH from 1955 to 1968, during which time a 15-fold increase in the budget (from $81 million to $1.2 billion) expanded the broad scientific base of medicine. Representative John E. Fogarty, a bricklayer from Providence, Rhode Island, and J. Lister Hill, Senator from Alabama, were the patron saints in Congress who made this growth possible.

Among the policy decisions, I would emphasize these:

(1) to expend most of the budget extramurally in grants to universities and private research organizations,

(2) to award these grants to individuals, young and old, rather than to departments or institutions,

(3) to make these awards purely on scientific merit as judged by a panel of peers drawn from outside the government,

(4) to be unswayed by political or geographical considerations, national boundaries included,

(5) to support basic research, even within the purview of each of the categorical disease institutes (Heart, Cancer, and so on),

(6) to provide fellowships and grants for training of pre- and postdoctoral students, and

(7) to enlarge the intramural resources (in Bethesda) to accommodate a large expansion of research and postdoctoral training.

Perhaps the most significant of these policies is grant support to individuals selected by peer review. In 1959 I was one of a group of five American biochemists (the others were Konrad E. Bloch, Herbert E. Carter, Bernard D. Davis, and the late Albert L. Lehninger) visiting the Soviet Union as part of an exchange program between our National Academy of Science and theirs. After a month of observing the management of research in the major Soviet universities and institutes, the Minister of Science asked us to compare the Soviet and American systems.

We said diplomatically: "Your system is different. You place authority for direction of research in the hands of a Director. In the United States, the individual scientist is in control. Immediately after completing his training, the young scientist applies for a research grant and is judged in competition with other applicants by a group of peers outside his institution, scientists within his special area of science. With the award of a grant, he becomes his own boss. His success or failure depends on what he accomplishes."

Our Russian host was puzzled: "It is your system that is different," he said. "Our system is the same as that practiced in all other countries, in Europe and Japan." He was right, and it is still true today that in most of the world, direction of research is vested in a relatively few senior people, whereas in the United States the bulk of research money in biology goes to thousands of individual investigators. Some complex problems may require the energies and disciplines of several individuals in a group effort. But far and away, award of research grants to individuals by peer review works best. Progress in science depends on the creative energies of the individual.

An aspect of the NIH grants program which deserves more notice is the award of grants for support of research outside the United States. From the very outset after World War II, when funds were inadequate to support all the qualified and deserving American scientists, the decision was made to award grants to the most outstanding scientists and laboratories, regardless of where they were in the world. I am pleased in having had a role in that decision.

The advantages of this international spirit in promoting science proved to be far greater than we expected. In addition, we had not anticipated the enormous boost this altruism gave to medical science and technology in the United States. By rejuvenating European and Japanese scientists and laboratories—in effect a "mini-Marshall" plan for sciences comparable to the massive postwar Marshall plan for economic rehabilitation—we were able to enlist the vast reservoirs of talent on all three continents. In so doing, knowledge was obtained which we could all share and thereby generate the most remarkable advance in medical science the world has ever known.

As a consequence of the rebirth of science centers in Europe and Japan, a tide of gifted students and senior investigators flowed into the United States. We welcomed them, and many remained to enrich American universities, research institutes, and industries. At the NIH laboratories in Bethesda alone, many thousands of foreign scientists (over 3,000 from Japan alone) received postdoctoral training and became loyal alumni upon returning to their native countries. These developments also helped to create markets for American technology and pharmaceutical products and to establish English as the international language of science.

Were the NIH record to be described for publication as an experiment in research administration, an impartial reviewer, even in this social area of science, might well question whether other factors might have been responsible for the good result. Such an experimental control does in fact exist in the support program of agricultural science in the United States during the same postwar period. The Department of Agriculture retained all authority within its own bureaucracy and limited research activity to its few established regional laboratories around the country. There were no grants to universities and private institutes. With this old-fashioned system of management, the knowledge base for agriculture remained stagnant. Little was learned about the basic biochemistry and genetics of plants and farm animals. Only recently, with the introduction of recombinant DNA technology, has there finally been a slight awakening of interest and activity in basic agricultural science.

The Alumni Reunion of 1975 was held not for sentimental reasons, nor to publicize past and present achievements, but rather to express the concern of alumni for the future of NIH. Despite its superb record and its dedication to science and the conquest of human disease, NIH had become and remains the target of budget cutters and antiscience forces. As with all worthwhile things, the struggle for survival is never won. This is even more true of support for science than of support for other institutions in society.

The difficulty with research support in our society, I have come to realize, is the failure to understand the nature and importance of basic research. This failure can be seen among members of the lay public, political leaders, physicians, and even scientists themselves. Most people are not prepared for the long time-scale of basic research and the need for a critical mass of collective effort. Fragments of knowledge unwelcomed and unexploited are lost, as were Gregor

Mendel's basic genetic discoveries. The vast majority of legislators cannot accept the seeming irrelevance of basic research. Were there a record of research grants in the Stone Age, it would likely show that major grants were awarded for proposals to build better stone axes and that critics of the time ridiculed a tiny grant to someone fooling around with bronze and iron. People do not realize that when it comes to arguing their case for more funding, scientists who do basic research are the least articulate, least organized, and least temperamentally equipped to justify what they are doing. In a society where selling is so important, where the medium is the message, these handicaps can spell extinction.

Placing Orotic Acid on the Main Track

In 1953 we knew that orotic acid could become a pyrimidine building block of nucleic acids, but we were not sure whether it was on the main track or connected by a spur. To find out whether orotic acid went straight "up" to a uracil nucleotide, we obtained orotic acid tagged with radioactive carbon in its carboxyl group, expecting that, were it converted to the uracil precursor of RNA, the carboxyl group would have to be released and could then be collected as radioactive carbon dioxide (Fig. 5-1). In fact, a tiny bit of radioactivity was released from the tagged orotic acid when we exposed it to extracts from yeast or liver. But the reaction was so very feeble. Less than one-tenth of one percent of the orotic acid molecules were affected. Yet, we were encouraged because this reaction, puny as it was, still required that we furnish the very component that would be needed to make uracil into a nucleotide building block, namely a ribose sugar with phosphate attached to carbon five. There was also a need for ATP, presumably to supply energy or another kind of activation. During the summer of 1953, while I was in Pacific Grove, California, learning general microbiology, Lieberman, who had come with me to Washington University as an instructor, remained in St. Louis, working hard but making no headway in boosting the efficiency of the reaction.

Upon returning to St. Louis that fall I had a lucky thought. Instead of using extracts of yeast or liver, why not put both together? Both yeast and liver cells make uracil from orotic acid, presumably by the same pathway. Perhaps our liver extracts were deficient in one of the enzymes and the yeast in another. Wow! The reaction was explosive, hundreds of times greater than before. The radioactivity counter went wild—one of those rare moments in a scientific lifetime. Evidently, at least two enzymes were indeed needed. As I had guessed, one of the enzymes was abundant in the yeast extracts but failed to survive in our liver extract preparation; the other enzyme was poorly extracted into our yeast preparation but was plentiful in the one from liver. When mixed, the liver and yeast extracts complemented each other. Now we would purify the two activities. "Lieb, which would you like to purify?" "I'll take the yeast enzyme," he said. That left me the liver. The liver enzyme proved to be the more novel and glamorous, a clear candidate to be one of the fairest of enzymes. Lieb often grumbled about my luck in getting the pretty one on this double-blind date.

I could show that, in the sequence of reactions, the liver extract produced something from two of the reactants (ATP and ribose phosphate) which could later be used by the yeast extract upon addition of orotic acid. Starting with powdered pigeon liver, I extracted a mixture of proteins that catalyzed the reaction between ATP and ribose phosphate 25 times better than the crude starting material. The product of this reaction enabled the yeast enzyme to release carbon dioxide from orotic acid (Fig. 5-5). Precisely what was

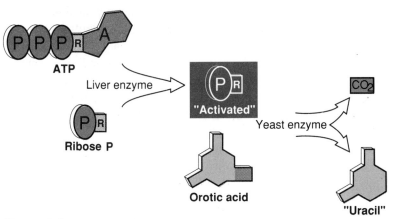

FIGURE 5-5
The liver enzyme uses ATP to produce an activated form of ribose-P, which the yeast enzyme then uses to release carbon dioxide from orotic acid to produce an unidentified form of uracil.

the liver enzyme doing? The best guess was that it activated the ribose phosphate by transferring the terminal phosphate of ATP to its number-one carbon, the one to which orotic acid would later be attached. In fact, there was a published claim in the British journal *Nature* of such a ribose derivative with two phosphate groups (one at carbon 1 and the other at carbon 5) and rumors that Buchanan's laboratory had found the same.

We would now try to separate and purify the activated product from the substrate we started with using the new and powerful technique of chromatography, to which I had been introduced four years earlier. The technique originated in 1906, in the work of a Russian botanist, Michael Tswett. He filtered an extract of leaves through a narrow glass tube packed with powdered chalk and was startled to see, "like the light rays in a spectrum," distinct bands of the different chlorophyll greens and carotenoid yellows, reds, and oranges appearing as liquid percolated through the column (Fig. 5-6). Less strongly adsorbed pigments migrated more rapidly and emerged from the column consecutively as relatively pure substances. "Such a preparation," wrote Tswett, "I term a chromatogram. It is self-evident that the adsorption phenomena described are not restricted to the chlorophyll pigments, and one must assume that all kinds of colored and colorless chemical compounds are subject to the same laws."

During the Manhattan project to build an atomic bomb, columns packed with ion-exchangers (resembling those now used to deionize water in the home) were used to separate and concentrate the trace products of nuclear fission. In 1949 I spent a day at Oak Ridge, Tennessee, with Waldo E. Cohn, who had learned ion-exchange chromatography from his war work and was now applying the method to separate nucleotides derived from the enzymatic breakdown of nucleic acids. I was thrilled and astonished by the power of this technique. A mixture of nucleotides, inseparable by previous methods in which metals were used for selective precipitation, could now be sorted out simply by passing them through Dowex-1, an ion-exchanger made by the Dow Chemical Company. The separation and recoveries were complete in a few hours. It was easy. It was beautiful. I immediately applied ion-exchange chromatography to discover exactly where the third phosphate of the NADP coenzyme was located. Now, in 1954, I would attempt to apply this method to the isolation of the unstable activated form of ribose phosphate.

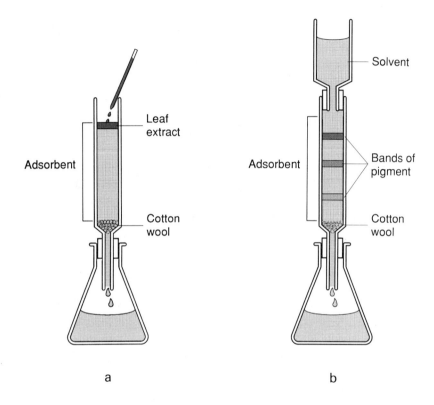

a

b

FIGURE 5-6
Column chromatography. A mixture of substances (such as pigments extracted from a leaf) is applied to a column packed with an adsorbent,
such as starch or chalk (a). If an appropriate solvent is then run through
the column (b), the substances are separated by virtue of the adsorbent.

Ernie Simms, my research assistant, first tried to separate the product of the liver enzyme action from unreacted ribose phosphate. As samples emerged serially from the ion-exchange column, a cluster of them ("a peak") contained most of the liver enzyme product that enabled the yeast enzyme to release carbon dioxide from orotic acid. He pooled these peak samples and determined their phosphate content. The ratio of phosphate to ribose, as we all anticipated, was indeed close to two. It confirmed the claims by others that activation of ribose phosphate entailed the addition of a single phosphate group to carbon 1.

Nevertheless, I was uneasy with the result. It was based on an analysis of a *pool* of samples. I insisted that we run more columns and analyze each sample as it emerged from the column, rather than pooling the samples. When we did this, the values we obtained were strikingly different. In each of the samples from the peak, the ratio of phosphates to ribose was nearer three than two. Working rapidly and at near zero degrees to counteract the presumed instability of the ribose phosphate product, we were able to isolate it intact. There were two phosphates (a pyrophosphate group) at carbon 1, the result of an entirely new kind of reaction! PRPP is the acronym we applied to 5-phosphoribosyl-1-pyrophosphate, the activated ribose phosphate; *PRPP synthetase* is the name we gave the enzyme that makes it.

Three years later, in 1957, Gobind Khorana, on one of his working visits to my lab, helped show how the synthetase makes PRPP from ATP. In the three-phosphate chain of ATP, the terminal pair is captured by a ribose phosphate attack on the middle phosphate (Fig. 5-7). During this stay in St. Louis, Khorana was repaying a visit my family and the Bergs (Millie and Paul) had made the previous summer to his lab in Vancouver, British Columbia. Paul and I had gone there to learn his novel chemical techniques for joining phosphates and amino acids to nucleotides in order to produce the precursors we needed for the enzymatic synthesis of DNA and proteins. Khorana in subsequent years, by a unique combination of talents in organic chemistry and enzymology, synthesized fragments of DNA and deciphered the genetic code, which earned him a share of the Nobel Prize for 1968.

I feel a kinship with Gobind despite our very different backgrounds. We both keep striving to make up for a late start in science. He grew up in the only literate family in a Punjabi village of one

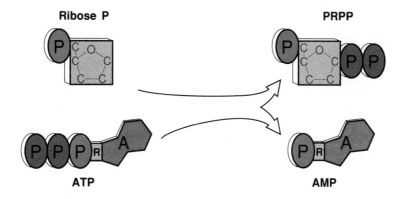

FIGURE 5-7
Ribose-P is activated by acquiring the terminal two phosphates of an ATP molecule to form PRPP (phosphoribosyl pyrophosphate).

hundred people. Monthly visits by an itinerant teacher were hardly the preparation he needed for a university curriculum. Yet he did well enough to get scholarship support to go on to graduate studies in Leeds and postdoctoral fellowships with Vladimir Prelog in Zurich and Alexander Todd in Cambridge. For lack of a job in India, he took a research post in Vancouver, but was ignored by the chemistry faculty at the University of British Columbia. Finally, at the University of Wisconsin in 1960, he had the resources to exploit his extraordinary abilities and energy to bridge the cultures of chemistry and biology.

Once when Gobind, Francis Crick, and I happened to be together, Crick asked us what prompted our working on DNA. Gobind answered that his success in the chemical synthesis of ATP led him to the synthesis of the more complex coenzyme A, which in turn led him to more and more difficult condensations of chains of nucleotides. Finally, this trail brought him to the synthesis of a stretch of DNA, a gene. With him, as with me, one thing had led to the next. We were aware of the importance of DNA, but we were as fascinated by the journey as by the destination. Crick's style, by contrast, has been to search for grand designs, and, to his great credit, he has found them.

I recall amusing incidents during the 1954 PRPP work which feature José Fernandes, a Rockefeller Foundation Fellow from the Uni-

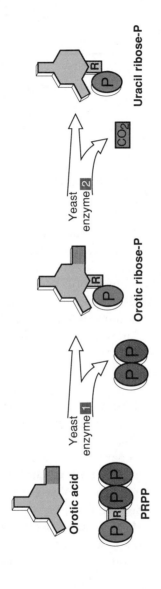

FIGURE 5-8

PRPP is directed by one yeast enzyme to transfer its activated ribose-P to orotic acid. The product, orotic ribose-P, then loses carbon dioxide through the action of another yeast enzyme to become uracil ribose-P (uridine monophosphate, UMP), a building block of RNA and a precursor of DNA.

versity of São Paulo and, in recent years, a successful citrus rancher. His laboratory notes would describe the companions who went with him to the brewery to obtain yeast, and the route they took, and would include such remarks as, "Oh my goodness! I threw away the precipitate instead of the supernatant." One day, while reviewing his experiments, I said: "José, I don't know what you're driving at." Excited and gesticulating, he responded: "Me driving? No! You drive! You drive all the time!"

With PRPP in hand, Lieb purified the yeast enzyme activity that removes carbon dioxide from orotic acid to produce the uracil nucleotide (Fig. 5-8) and found that two separate enzymes were needed. The first directed orotic acid to attach itself to carbon 1 of PRPP by displacing the pyrophosphate group and releasing the now familiar inorganic pyrophosphate. The second enzyme then removed carbon dioxide to produce uracil ribose phosphate, the uracil nucleotide found in RNA.

Alternate Pathways to the Building Blocks of DNA

PRPP had unlocked the door to the assembly of uridine (uracil ribose) phosphate, one of the four nucleotides that constitute RNA. Would it also be a key to how the other nucleotides of RNA and DNA are made? Was there an activation of these several nucleotides, like that for adenosine (adenine ribose) monophosphate, which with successive phosphate additions becomes a diphosphate and then a triphosphate, the vaunted ATP? These and other exciting possibilities filled my mind to overflowing—more things to grasp than I could hold. I felt uneasy and anxious that precious discoveries would slip away.

There were only five in my research team in 1954. Two of the five—my wife Sylvy and Paul Berg, who had just arrived as a postdoctoral fellow—were engaged in other projects that were also exciting. Sylvy had the scent of the enzyme that makes the mysterious polyphosphate, a long chain of phosphates. Massive accumulations in yeast and other cells were once called volutin and mistaken for nucleic acid with its phosphate-rich backbone. Since its phosphates are activated, as in ATP or pyrophosphate, the metabolic importance of polyphosphate might rival that of ATP.

Paul was discovering new sources of pyrophosphate. When he arrived the previous year, I had wanted to know whether the synthesis of coenzymes could account for all pyrophosphate production. We had a simple assay to find out. When coenzyme synthesis is reversed, radioactively labeled pyrophosphate enters ATP. The ATP can then be totally adsorbed on powdered charcoal. In this way, replacement of the terminal phosphates of ATP with pyrophosphate could be measured easily, even in a crude cell extract. (For many years, a good batch of charcoal was treasured as "black gold.") Unexpectedly, he found that such exchanges were also effected by other enzymes, one of which used ATP to activate acetic acid, another to activate the amino acid methionine. These and like reactions later proved to be key events in the assembly of membrane lipids, steroid hormones, and proteins.

Since Sylvy and Paul could not help, Lieb, Simms, and I would have to work even harder to gather the harvest of nucleotide discoveries. We promptly found enzymes in extracts of liver and yeast that converted uridine monophosphate to the di- and then the triphosphate forms. This, in turn, enabled Lieb to find another enzyme that used uridine triphosphate as a substrate; attachment of ammonia (NH_3) produced cytidine (cytosine ribose) triphosphate, the activated state of the other pyrimidine component of RNA. We also found an enzyme that condensed PRPP with adenine to form adenine ribose phosphate (ARP), the adenine nucleotide, and then another enzyme (HGPRTase) that did the same with guanine to form the other purine nucleotide. Now it seemed we had the four building blocks of RNA and should press on to find the enzymes that assembled them into the giant molecule.

Puzzling at first was the fact that the pathway to the pyrimidine nucleotides differs from that to the purines. To make the uracil or cytosine nucleotides, the pyrimidine ring (orotic acid) is assembled first and then attached (via PRPP) to the sugar phosphate. In the synthesis of the purine nucleotides (adenine and guanine), as Buchanan and Greenberg discovered, an activated ribose phosphate (later identified as PRPP) provides the foundation upon which the complex purine ring is built from smaller units.

Still, we were now finding enzymes that could also condense preformed adenine or guanine with PRPP to make the nucleotides directly. Apparently, cells had long ago evolved the ability to exploit something as precious as an already formed purine ring, if it hap-

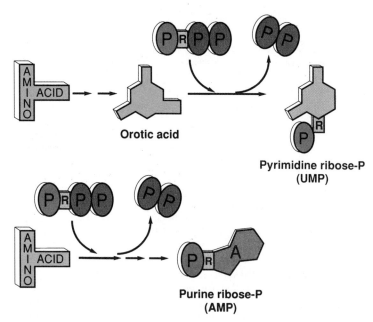

Orotic acid

**Pyrimidine ribose-P
(UMP)**

**Purine ribose-P
(AMP)**

FIGURE 5-9
Biosynthesis of pyrimidine nucleotides (such as UMP) entails the introduction of ribose-P (from PRPP) *after* the pyrimidine structure (for example, orotic acid) has been formed. By contrast, in the biosynthesis of purine nucleotides (such as AMP), PRPP is incorporated *at the very outset* in building the purine structure from smaller molecules (such as amino acids).

pened by in the surrounding medium. Did cells have comparable mechanisms to salvage each of the purines and pyrimidines with an attached sugar but lacking phosphate? Indeed. We found enzymes, called kinases (Greek *to move*), that transferred phosphate from ATP and made nucleotides from a precursor that lacked phosphate.

It became clear to me that cells could make use of two pathways to produce the building blocks of the nucleic acids. By one pathway, which I called *de novo* (Latin *anew*), the building block is laboriously put together from parts of simpler molecules (sugar phosphate, amino acids, carbon dioxide) in a sequence of ten or more enzyme-directed reactions (Fig. 5-9). The building blocks adenine (A), guanine (G), cytosine (C), uracil (U), and thymine (T) are produced with a sugar phosphate attached, and in this activated state are ready for assembly into nucleic acids. By another set of pathways, which I

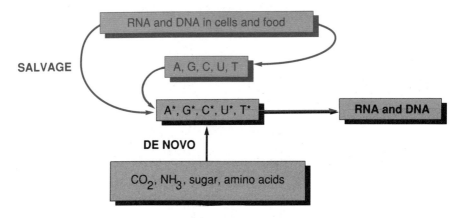

FIGURE 5-10
Pathways of the biosynthesis of nucleotides are of two general kinds: *de novo pathways,* in which the nucleotides (A*, G*, C*, U*, T*) are built up from smaller molecules (carbon dioxide, ammonia, sugar, amino acids); and *salvage pathways,* in which the nucleotides or parts of them (A, G, C, U, T) are obtained from the breakdown of RNA and DNA in the digestion of food or their turnover in cells.

called *salvage,* parts of building blocks released by wear and tear of nucleic acids in cells, or by their generation during digestion of the nucleic acids in food, are recycled to produce the activated building blocks (Fig. 5-10).

What I knew about the biosynthetic routes to the nucleic acid precursors was based on enzymatic studies with bacteria, yeast, and pigeons. I had mistakenly assumed that traffic on the salvage route would be minor compared with that on the *de novo* pathway. Also, the denigratory tone of the word "salvage" diverted us from appreciating the vital importance of these pathways. I was not attentive enough to human disorders of purine synthesis, such as gout, of which bacteria and pigeons do not complain. I was therefore surprised in 1967 to learn that the ghastly and fatal Lesch-Nyhan syndrome could be caused by a deficiency in a particular salvage enzyme, HGPRTase. Children born with this mutation, a defect in the enzyme that salvages G (by adding sugar phosphate to it), convert the G to uric acid and end up with severe gout and profound mental aberrations (see Chapter 3). Additional examples were discovered in which the defective function of other salvage enzymes result in a backup of metabolic traffic that leads to deficiencies in the immune system and other fatal disorders.

When quantitative measurements were made on humans as well as on microorganisms, we came to recognize that salvage pathways are essential in many circumstances. Brain cells, which lack the *de novo* pathway, rely totally on salvaging precursors produced by liver cells. Organisms and cells deficient in a *de novo* pathway as a result of disease, poisons, drugs, and metabolic stress, or an inadequate rate of building-block production for any reason, rely on salvage pathways. Recycling is very important in metabolism. When it fails, a massive accumulation, as in the case of uric acid, becomes a serious disposal problem. The patterns of molecular traffic in the synthesis, disassembly, and salvage of building blocks are intricate and ingeniously controlled. These patterns, while basically the same throughout nature, differ in details among species and among various cells and tissues within an organism.

For some twenty-five years in lectures on nucleic acid metabolism, I described gout, a disease unique to humans, as the result of two evolutionary misfortunes. One was the loss by all primates and birds of the gene for an enzyme, uricase, which in other mammals and lower forms converts uric acid to allantoin, a far more soluble substance that is easily excreted. The other was the acquisition of a renal mechanism, lacking in other primates, that reabsorbs more uric acid back into the bloodstream. As a consequence of these two mutations, uric acid is maintained in human blood and tissue fluids at a level ten times that of other primates—one poised near the point of crystallization and trouble.

Now it seems I may have been mistaken in thinking that the evolutionary selection of mechanisms that retain uric acid in humans at the highest possible level was a misfortune. Bruce Ames at the University of California at Berkeley points out that uric acid, a potent antioxidant, can serve as a first line of defense in the detoxification of substances in food and in the atmosphere that are implicated as carcinogens, mutagens, and contributors to the aging process. In this role, it may be even more effective than megadosage of vitamin C at two grams a day. Rather than being an error, maximal levels of uric acid in human body fluids would appear instead to be an effective adaptation over a 60-million-year period of primate evolution to compensate for loss of the capacity to synthesize vitamin C and to make up for a dietary switch away from huge amounts of vitamin-C-rich vegetation.

Synthesis of DNA in a Broken-Cell Extract

One of the spurs to embark on the synthesis of nucleic acids came in 1953 as a result of discharging an obligation I was feeling as the newly ordained chairman of Washington University's medical microbiology department in St. Louis. In trying to introduce some biochemical flavor into the traditionally descriptive microbiology curriculum, I designed a class laboratory exercise demonstrating that certain bacteria (such as streptococci) secrete enzymes that degrade DNA. I isolated some DNA for this exercise from calf thymus glands. In the final step of the isolation procedure, the white threads of the genetic material are wound on a rod into a tuft like transparent cotton candy. (Seeing DNA directly was almost as thrilling as a new discovery. Few things in biochemistry are visually exciting; usually, we can only follow the movement of a spectrophotometer needle or watch the flashing of a radioactivity counter as signals of success or failure.) Having prepared a generous supply of DNA for class use, I was able to divert some of it for experiments on DNA synthesis that I would otherwise have put off.

Having learned how the likely nucleotide building blocks of nucleic acids are synthesized and activated in cells, I naturally turned in 1954 to the search for enzymes that assemble these nucleotides into RNA and DNA. Such a quest was considered by most of my colleagues to be audacious. Synthesis of starch and fat, once regarded as impossible outside the living cell, had been achieved with enzymes in the test tube. Still, the monotonous array of sugar units in starch or the acetic acid units in fat was a far cry from the assembly of DNA, thousands of times larger and genetically precise.

Yet all I would be trying to do was follow in the classical biochemical traditions of this century as practiced by my teachers. It always seemed to me that a biochemist devoted to enzymes could, if persistent, reconstitute *any* metabolic event in the test tube as well as the cell does it. In fact, better! Without the constraints under which an intact cell must operate, the biochemist can manipulate the concentrations of substrates and enzymes and arrange the medium around them to favor the reaction of his choice.

How fortunate that molecules of RNA or DNA clump together (precipitate) when a solution is acidified, while at the same time the nucleotides from which they are made remain soluble. Because of this simple fact, nucleic acids can be easily freed of nucleotides and collected. By tagging a nucleotide with radioactive phosphorus or carbon, one can trace the incorporation of relatively few nucleotides into an RNA or DNA chain. This is the approach we used with extracts from the common intestinal bacterium *Escherichia coli*. By 1954 *E. coli* had become the favored object of biochemical and genetic studies worldwide, replacing yeast as our preferred source of enzymes from a rapidly growing cell.

I explored the synthesis of RNA with my new postdoctoral fellow, Uri Littauer. Uri had come to my lab from the Weizmann Institute in Israel. A native Israeli, he had fought in the 1948 War of Independence and was among the first of the soldier-scientists to do his doctoral studies in Israel. He later returned to the Weizmann Institute to become one of the leaders of this outstanding research center. Uri and I prepared ATP, an activated form of the adenine (A) nucleotide, labeled in the adenine. Uri observed that after ATP was exposed to an *E. coli* extract, a small but significant amount of its radioactivity could be made precipitable (presumably RNA) upon acidification.

Were we observing a genuine enzymatic reaction? We inferred that enzymes were responsible for this conversion for three reasons: (1) The activity in the cell extract was unstable, (2) the extent of the reaction depended on time (10 to 30 minutes), and (3) the optimal temperature was in the range of 30 to 37°C, favored for growth of *E. coli*. Littauer was making progress in fractionating the proteins in the extract to enrich for those needed in the reaction and to remove irrelevant and interfering activities. We felt we were onto something very important, the enzymatic synthesis of RNA.

In 1955 I was also pursuing the synthesis of DNA. Here, I had the invaluable help of Morris Friedkin in a nearby building at Washington University. He had joined the Department of Pharmacology after a postdoctoral fellowship with Herman Kalckar in Copenhagen, where he had found an enzyme that makes thymidine. DNA is distinguished from RNA by having thymine (T) in place of uracil (U) and by having the deoxyribose sugar, which lacks the oxygen present at carbon 2 in ribose (Fig. 5-11). Friedkin's enzyme condenses thymine with deoxyribose 1-phosphate (Fig. 5-12) in a reaction analogous to one that Kalckar had discovered for purines and pyrimidines reacting with ribose 1-phosphate.

Friedkin had been given a small amount of thymine (labeled with radioactive carbon 14), to which he attached deoxyribose to make

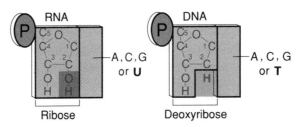

FIGURE 5-11
Two features distinguish RNA from DNA: ribose and uracil
(U) in RNA and deoxyribose (lacking an oxygen at carbon
2) and thymine (T) in DNA.

thymidine, using his enzyme. He showed that the tagged thymidine
is taken up by the rapidly growing cells in rabbit bone marrow or
onion root tips and built into the DNA of these tissues. Despite my
urging, Friedkin was not inclined to search for incorporation of thymidine into DNA with extracts from broken cells. But he generously
saved the reaction liquid remaining after his cellular experiments.
From these discards, I recovered radioactive thymidine to use in
trials with extracts of E. coli.

The results were mixed. Unlike Friedkin's success with bone marrow and onion roots, very little of the radioactive thymidine was
converted to a form with the properties of DNA. After one hour of
interaction with the bacterial cell extract at 37°C, a tiny amount of
radioactivity was trapped in a pellet, insoluble in acid, and possibly
indicative of DNA. The amount measured 68 disintegrations per
minute, less than 0.01 percent of the one million disintegrations per
minute in the thymidine we started with. This value was only
slightly above that found in an unreacted sample of thymidine, one
exposed to a cell extract kept at 0°C. The increment attributable to
enzyme action was at the border of significance. Because results with
further trials were no better, I shifted my attention to a more vigorous
reaction. From 5 to 10 percent of the thymidine was converted to
novel soluble forms that resembled the phosphorylated states of the
nucleotide building blocks. We called these new compounds "thymidine X" and hoped they might provide better precursors than
thymidine for DNA synthesis.

At this juncture, in the spring of 1955, Herman Kalckar came to St.
Louis for a visit. He brought us some startling and unsettling news.

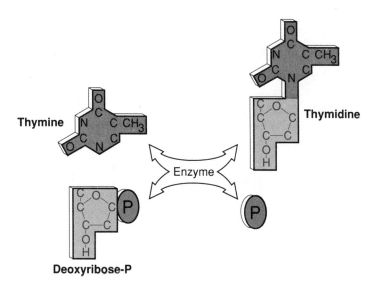

FIGURE 5-12
Thymidine, a precursor of DNA, can be synthesized by an enzyme that directs thymine to attack deoxyribose-P (activated at carbon 1), releasing the phosphate. The reaction is reversible.

For Littauer, it was devastating. Ochoa and his colleagues at New York University had just discovered the enzymatic synthesis of RNA from ADP (adenosine diphosphate). It was for them a totally unexpected finding. Marianne Grunberg-Manago, a postdoctoral fellow, had been studying mechanisms of ATP synthesis in energy metabolism using extracts from a nitrogen-fixing bacterium (*Azotobacter vinelandii*). Upon discovering what appeared to be an incorporation of radioactive phosphate into the terminal phosphate of ATP, she later found that the phosphate was instead entering ADP molecules that had contaminated the ATP preparation. Then, having purified the responsible enzyme with Ochoa's urging and direction, she became puzzled at being unable to account for much of the starting ADP. ADP turned up among compounds which resembled RNA, and she then showed that the enzyme made RNA-like chains by condensing many molecules of ADP or the other nucleoside diphosphates (such as GDP, UDP, CDP) (Fig. 5-13).

The reaction was readily reversible; cleavage of RNA by phosphate converted it to nucleoside diphosphates. Ochoa named the enzyme polynucleotide (RNA) phosphorylase in recognition of its

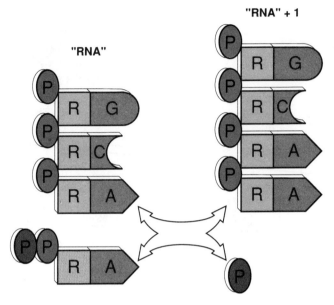

FIGURE 5-13
An RNA-like chain can be assembled from adenosine diphosphate (ADP) or other nucleoside diphosphates with the release of a phosphate group. The reaction, which is reversible, is catalyzed by an enzyme called polynucleotide phosphorylase.

capacity to cleave RNA with phosphate as well as to make RNA. The name was a prescient choice, in that subsequent research showed degradation, rather than synthesis, of RNA to be the true cellular function of the enzyme.

On the strength of this new information, we shifted to using ADP in place of ATP in our studies with *E. coli*. Initially, we had taken pains to exclude ADP, by using a potent "regenerating system": a large supply of a phosphorylated compound along with a specific enzyme that rapidly transfers its phosphate to ADP and thereby converts every trace of ADP to ATP. With ADP minus the regenerating system, the rate and extent of our reaction was greater, and we made rapid progress in obtaining a highly enriched enzyme from *E. coli*; it had properties much like those described by the Ochoa group.

We were keenly disappointed. We had merely confirmed the existence of polynucleotide phosphorylase in another bacterium. In so doing, we made a classic blunder and missed the discovery of the true RNA synthetic system. Being able to account for a phenomenon does not ensure that it is the only or best explanation of it. In this instance, we were diverted from the far more important discovery of the RNA polymerase which depends on ATP rather than ADP. By switching to ADP and tracking the synthetic activity of polynucleotide phosphorylase, we missed the key enzyme of cell growth and function, the enzyme responsible for transcribing genes on the route to synthesizing proteins.

Nearly a year passed before I repeated the experiment of trying to convert radioactive thymidine to an acid-insoluble form, indicative of DNA. The results in December 1955 were the same: Once again, only a tiny amount of this presumed precursor was converted. But two things had changed. For one, the radioactivity of the thymidine happened to be three times as great and so the results seemed more impressive. For another, believing I had lost out on the synthesis of RNA, I began to view the synthesis of DNA as an even more precious prize.

It was a simple matter to test the presumption that the radioactive thymidine had been incorporated into a DNA molecule. I exposed the product to an enzyme from the pancreas called DNase (deoxyribonucleic acid-ase), whose function in digestion is to break down the giant DNA molecule into its constituent nucleotides. After a few minutes of DNase action, radioactivity which had previously been trapped in the pellet after acid treatment was no longer there. It really looked as though the conversion of thymidine, puny though it was, had made it part of DNA.

Without the encouragement that the diagnostic action of DNase gave me, I wonder whether I would have had the will to pursue such a feeble light. Proper homage has never been paid to this member of the enzyme proletariat, nor to those who made it available. The enzyme, isolated from pancreatic juice, degrades only DNA. Maclyn McCarty and Oswald T. Avery at the Rockefeller Institute first used the enzyme to prove, in one of the most historic experiments in biology, that DNA is the genetic material. Moses Kunitz later crystallized the enzyme.

Even before I calculated the results from the radioactivity readings on the Geiger counter, I stopped to tell Robert Lehman about the DNase experiment. Bob had arrived three months earlier to work with me as a postdoctoral fellow. He had been an infantryman in Europe during World War II and had returned to Baltimore to attend Johns Hopkins University for his college and graduate degrees. In

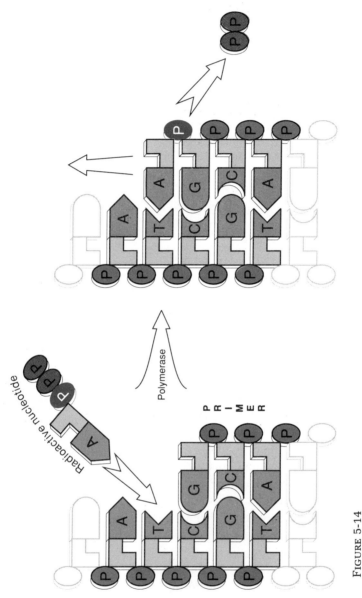

FIGURE 5-14

DNA synthesis, catalyzed by DNA polymerase, requires three elements: (a) a primer chain, to provide a growing end to which the nucleotide is added; (2) a nucleotide (activated as a triphosphate) that matches the next open place on the template; and (3) a template, to direct the incoming nucleotide by the matching of A to T or G to C.

only a few weeks Bob had made an impressive start on an important problem. DNA of the T2 virus that infects *E. coli* is unlike other DNA in having its cytosine (one of the four DNA constituents) modified by having a CH_2OH group attached. We had wondered how this novel form might be made. Bob had already found extracts of cells infected with the T2 virus that could effect this modification of a cytosine nucleotide, and he was beginning to purify the enzyme responsible for this event.

I was astonished that, on hearing my DNase result, Bob said instantly he would like to drop his project and work with me on DNA synthesis. What a bold and fateful decision! The very next day, Bob moved from his laboratory room into mine. He confirmed my results and in succeeding days boosted the level of DNA synthesis by a series of experiments that convinced us we were onto something complicated and important. Bob soon found that thymidine phosphate, the whole nucleotide, was a far better precursor than thymidine and increased synthesis two- to three-fold. Later, he showed that thymidine triphosphate was much better still.

With improvements in the assay of DNA synthesis by these crude extracts, our goal became clear: to purify the enzyme that assembled nucleotides into a DNA chain—the DNA polymerase. The most complex and revealing insights into the reaction would come from exploring the function of the DNA that I had included in the reaction mixture in my earliest attempt to incorporate thymidine into DNA.

In accounts I have since heard about the discovery of DNA polymerase, it is said that I included the DNA to serve as a template because of the role proposed for it two years earlier in the Watson and Crick model for replication. The DNA could come from any source: *E. coli*, calf thymus gland, or salmon sperm. By these accounts, the idea of a primer role for DNA—providing chain ends for extension (Fig. 5-14)—emerged many years later.

Not so. In the earliest experiments, I added DNA expecting that it would serve as a primer for growth of a DNA chain. I did this because I was influenced by the work of Carl and Gerty Cori on the growth of carbohydrate chains by glycogen phosphorylase. They were the first to show that assembly of a polymer, in this case a starch-like chain, depends on having a preformed chain to elongate. I never presumed, in exploring the synthesis of DNA, that I would immediately discover a phenomenon so utterly unprecedented in biochemistry: one in which an enzyme is absolutely dependent on its substrate, a template, for instruction.

I had added DNA for a second important reason. Knowing that the breakdown of DNA by nucleases in extracts of disrupted cells far exceeds its synthesis, I hoped that some of the DNA molecules in the large starting pool would incorporate tagged thymidine by synthesis and that some would then survive the nucleases. Only after several months did Bob Lehman and I learn, with surprise and elation, that the added DNA, beyond shielding the product from nucleases and appearing to be a primer, fulfilled two other essential functions. It was indeed essential as a template and also provided the nucleotide building blocks we were missing. DNA accomplished this by being broken down into nucleotides by DNases in the extract. The nucleotides derived from its cleavage were then activated by five other enzymes in the extract, which used the ATP we furnished to generate the triphosphate forms of the A, G, and C nucleotides as well as the tagged T that we had added. At that time, these activated nucleotides were still unknown and could not have been added deliberately to our experimental brew.

The road was clear to proceed with the discovery and purification of all the enzymes essential for DNA synthesis, most of all, the polymerizing enzyme itself—DNA polymerase.

A Remarkable Copying Machine

Among my many love affairs with enzymes, the one with DNA polymerase has been by far the strongest and longest. Yet I retain sentimental attachments to many others. In May 1988, when my former and present students and colleagues gathered in San Francisco for a gala 70th birthday party, I thought it would be fun to select, from among the thirty or more enzymes I had worked with, the ten I favored most. I was at first surprised by the list. Six of the top ten were discovered between 1948 and 1955; that left room for only four to be selected from more than twenty attractive candidates which appeared in the next thirty years. (The six were: potato nucleotide pyrophosphatase, NAD synthetase, phosphatidic acid synthetase, PRPP synthetase, polyphosphate synthetase, and DNA polymerase.) Upon reflection, the basis on which I had made these choices became clear. I am sentimentally most attached to those enzymes that came during the time of my life when I collected the data myself, from conception to delivery.

Going over the laboratory notes of these early explorations for the synthesis of DNA in broken-cell extracts, I shudder to see the odds we faced and our innocence of them. The signal we followed was barely detectable, our supplies of the building blocks were scanty and primitive, the techniques for fractionating enzymes were few and blunt, and there were no genetic guides for enzymes in the synthetic pathway. Yet, my faith in the power of biochemistry was so great that, having devised an assay, I believed I could force a wedge into this tiny crack and use the hammer of enzyme purification to drive it to a solution.

In the spring of 1956 Maurice (Moish) Bessman, a postdoctoral fellow, and Steven Zimmerman, a graduate student, joined Bob Lehman, Sylvy, Ernie Simms, and me and one or two others in our twenty-by-twenty-foot laboratory. Julius Adler came as a postdoctoral fellow a year later. Crowded, excited, sharing good spirits, ideas, and reagents, we produced results the sum of which was far greater than it would have been were we diluted into a larger space or separated by walls.

By May, six months after we started working on DNA synthesis in cell extracts, I could report significant progress at a historic Johns Hopkins University symposium on "The Chemical Basis of Heredity." Among the thirty-seven papers were many of the tremors that would generate the tsunami of molecular biology. A student today, armed with our current knowledge of DNA polymerase actions, would be surprised by the uncertainties in my report. Yet it was clear that we were exuberantly on our way to understanding the replication of DNA.

The large molecules required for the synthetic reaction, separated into three fractions, were described as: a "heat-stable, DNA-containing fraction, at present regarded as a 'primer', and two enzyme fractions, S and P, each of which has been purified more than 100-fold." Fraction S possessed the enzyme that acted on the DNA fraction to generate the three missing nucleoside triphosphates (G, A, and C); fraction P contained the enzyme which assembled these into a DNA chain and which we began to call "polymerase." The enzyme, we said, "does so only under the remarkable condition that all four of the deoxynucleoside triphosphates be present." Growth of chains requiring all four building blocks of DNA must be directed by a DNA template, as proposed by Watson and Crick in their model. But we lacked the evidence to claim it.

Among our conclusions: "The overriding question remains: How is biologically specific DNA formed? The enzymatic synthesis of a bacterial transforming factor, once regarded beyond experimental reach, has now become an immediate objective." The "bacterial transforming factor" whose faithful replication we sought is a segment of DNA containing a gene responsible for a certain characteristic. Upon introduction into a second strain, this piece of DNA endows the recipient strain with the feature found in the donor strain. For assays of transforming factors, I went to the Rockefeller Institute in New York to get the advice of Rollin Hotchkiss for the factor in *Pneumococcus pneumoniae*; I invited Sol Goodgal of Johns Hopkins to St. Louis to show us how to assay for the *Hemophilus influenzae* factor; later at Stanford, Walter Bodmer and Joshua Lederberg helped us look for synthesis of genetically active *Bacillus subtilis* DNA. We were so hopeful!

First, we had to purify fraction P, the polymerase. We were raring to go! Once, many years later, when leaving the lecture hall in Berkeley after giving a seminar on my work, I overheard one student say to another: "How dull it must be to purify enzymes." It saddened me that I had failed to convey the excitement we all felt about this work. Perhaps I should have tried the mountain-climbing metaphor.

Choosing an abundant source of an enzyme is crucial before embarking on its isolation. A growing culture of *E. coli* proved to be best for DNA polymerase. Although we found polymerase activity in available animal tissues, their slow growth, only a hundredth as rapid as *E. coli*, was matched by correspondingly low levels of the enzyme. Virally infected *E. coli*, which make DNA at ten times the normal rate and had first attracted my attention to the mysteries of DNA synthesis, have complexities in their polymerase activities which took us several years to resolve. Later on, we examined extracts of another rapidly growing bacterium (*Bacillus subtilis*) only to find them teeming with enzymes that degrade proteins, including polymerase.

A significant fraction of our early efforts went into growing the bacteria. To get an ounce of bacterial cells, we grew them in 1 liter (about 1 quart) of culture medium in each of twelve 4-liter flasks aerated on a mechanical shaker at 37°C. It was a long and tiring day's work. Later, we acquired a 100-liter vessel in which 60 liters of a well-aerated culture yielded nearly a pound of cells. Eventually, we were able to purchase adequate supplies from the Grain Processing Corporation in Muscatine, Iowa. In a 10,000-gallon vat, normally used to make alcohol from corn steep liquor, they produced 200 pounds of bacteria for us. From this amount of starting material, it took us a month to obtain about one-half gram of nearly pure enzyme. It's easier now. With the advent of genetic engineering, we can program the *E. coli* cell to produce 100 times more of this enzyme and isolate it with less than one-tenth the effort; one pound of cells can yield a gram of pure enzyme in two days.

How to extract the juice of bacterial cells so minute that a billion of them are no larger than a grain of sand? We tried several methods of mechanical disruption: grinding with tiny glass beads in a mortar or a motorized blender; compression in a chamber at 2,000 pounds per square inch followed by rapid release through a narrow orifice; and bombardment by high-frequency sound waves. Sonic extraction gave us the best yield of DNA polymerase with the least effort and a reasonably stable starting material for isolation of the enzyme. Some years later, in searching for missing enzymes of DNA replication, these procedures proved too harsh, and we had to resort to more subtle chemical means to rupture the cells' outer walls and inner membranes.

In 1956, to purify DNA polymerase from extracts of *E. coli*, we tried a number of procedures that separate proteins according to size, shape, electric charge, and stickiness to various adsorbents. Because we did not know beforehand the properties of the enzyme, our choice of procedures had to be largely empirical. A good step yields 80 percent of the enzyme activity in a fraction that contains only 20 percent of the starting proteins, a 4-fold purification. One of the steps, a highly selective precipitation of nucleic-acid bound proteins with the basic antibiotic streptomycin, was sensational: a 40-fold purification with practically no loss of activity. After six steps, our enzyme was nearly pure: having started with one part in 5,000 in a mixture of thousands of different proteins in the crude extract, we had recovered about 20 percent of the original activity.

The immediate reward for purifying DNA polymerase was being rid of most of the enzymes that altered our substrates, particularly the nucleases that were degrading the DNA. Quantities of DNA were synthesized well beyond those we started with. They had the size, shape, and chemical properties characteristic of DNA isolated from animal, plant, or bacterial cells. Synthesis absolutely required the addition of DNA and the activated (triphosphate) forms of its A, G, T, and C constituents. Omitting any one of these reduced DNA synthesis more than 100-fold, down to the background levels—the levels observed without added enzyme or incubation at 30 to 37°C.

We described these findings in two papers entitled "Enzymatic Synthesis of Deoxyribonucleic Acid" and subtitled "I. Preparation of Substrates and Partial Purification of an Enzyme from *E. coli*" and "II. General Properties of the Reaction." They were submitted to the *Journal of Biological Chemistry* in October 1957. A month or so later the manuscripts were returned as unacceptable. There were no objections to the substance of the papers—a description for the first time of the novel triphosphate forms of the DNA building blocks, procedures for their synthesis, and the isolation of an enzyme that polymerized them. Nor were there criticisms of our characterization of the product—except that it should not be called DNA.

Among the anonymous referees, who eventually included ten leaders in nucleic acid research, some insisted that we call our synthetic product by the accurate but vapid name "polydeoxyribonucleotide" rather than DNA. One referee, whose caustic literary style unmistakably identified him to me, stated that we had to show genetic activity in our synthetic substance to have it qualify as DNA. Why? This criterion was met by less than 2 percent of the papers on DNA appearing in biochemical journals. After several exchanges of correspondence, I decided to withdraw the papers. At this juncture John Edsall, who was aware of the controversy and was to assume the editorship of the *Journal* in a month, stepped in. He wanted the papers published and asked me to wait. The papers appeared promptly, in the May 1958 issue.

Could we show that the DNA synthesized by the enzyme was a faithful copy of the template? For this to be true, the composition of the synthetic product would have to fulfill two criteria. One would be a characteristic found in all DNAs. Since A always pairs with T in the DNA duplex, the amounts of A and T should be equal; similarly the matching of G to C dictates the equivalence of these two. The other criterion would be a characteristic that distinguishes the DNA of the particular species used as a template, namely, the ratio of A-T pairs to G-C pairs; among various species, this ratio ranges from less than 0.5 to near 2.0.

In the summer of 1958 we were able to determine the composition of synthetic DNAs made with a variety of templates. The results seemed too good to be true, but there it was! The equivalences of A to T and of G to C were right on the mark: for five DNA primers, the values ranged from 0.98 to 1.02. As for the relative amounts of the four constituents, expressed as the ratio of A-T pairs to G-C pairs, the results were equally impressive. For example, when we started with DNA from *E. coli*, in which this ratio (A plus T over G plus C) measured 0.97, our product had a value of 1.02; with the DNA of T2 virus infecting *E. coli*, whose ratio was 1.90, our product ratio was 1.92. These fits were very close and gratifying. Furthermore, these ratios were unaffected even when we distorted the relative concentrations of the A, T, G, and C building blocks to start with or varied the extent of synthesis from 2 percent of the input template to over ten-fold. In the December 1958 issue of the *Proceedings of the National Academy of Sciences,* our report concluded: "These results suggest that the enzymatic synthesis of DNA by the 'polymerase' of *E. coli* represents the replication of a DNA template."

Were others convinced that the polymerase depended on guidance by a template? Despite these remarkable data, some biochemists were not. When I presented these results in a seminar at Yale University's Department of Biochemistry, my host, Joseph Fruton—a leading biochemist, author of the best biochemistry textbook of that time, and an erudite historian of science—was skeptical. "Arthur," he said, "an enzyme is designed to direct a specific reaction. There is no known case in which an enzyme takes instructions from its substrate." "Joe," I replied, "replication of DNA is unique in the life of a cell. Besides, how else can you explain my data?" There was no ready alternative, and there would be none. Still, many questions remained. Just how does an enzyme hold on to a template? How rapidly and extensively is the template copied? How accurately?

How accurately is the template copied? Mismatching a G with a T or misplacing an A opposite a C is a *mutation*, a mistake in the DNA that is likely to remain indelible. If a protein is encoded by that segment of DNA, it has a chance of being defective as a result of this replication error in the gene. How faithful was our enzyme in replicating the precise sequence of a gene, the average length of which is 1,000 nucleotides? Analyses of our synthetic DNA told us only that, within a few percent, we could reproduce the overall A, T, G, and C composition of DNA, but it revealed nothing about their sequential arrangement in our product. Current methods that readily determine the sequence of nucleotides in DNA were still more than two decades away in 1958.

With John Josse, a postdoctoral fellow fresh from a residency in medicine at Massachusetts General Hospital, I devised a procedure for determining the frequency with which one of the four nucleotides in a synthetic DNA is next to any other. There are sixteen possible "nearest-neighbor" sequences: AA, AG, AT, AC, GA, GG, GT, GC, TA, TG, TT, TC, CA, CG, CT, CC. The procedure for determining the percentage of the total represented by any one pair rests on several simple successive manipulations:

(1) DNA is synthesized with one of the four substrates (such as AÞPP) radioactively labeled in the innermost phosphate that will become part of the new DNA (Fig. 5-15).

(2) The synthetic product is degraded all the way to individual nucleotides by a nuclease (DNase) which splits the phosphate-sugar backbone in a particular way (Fig. 5-15). The split is invariably on the side of the labeled phosphate that separates it from the sugar to which it was linked in the original substrate. Thus, the labeled phosphate is left attached to the sugar at the end of the growing chain with which it reacted, its "nearest neighbor." In the example in Figure 5-15, the nearest neighbor of the radioactively labeled A substrate is G.

(3) The A, G, C, and T nucleotides are separated from each other by chromatography; the radioactivity measured in each indicates the frequency with which A was attached to A, T, G, and C.

(4) The entire procedure is repeated three times, with the radioactively labeled substrate being successively GÞPP, TÞPP, and CÞPP.

(5) The frequency or percentage of each of the sixteen arrangements of the four building blocks is easily calculated.

The results were all we could have wished. For each synthetic DNA there was a distinctive distribution of nearest-neighbor sequences. We could reasonably infer that this distribution reflected the sequences in the particular DNA which the enzyme had been given to copy.

Then, an even more impressive insight emerged from inspection of these data, one revealing a crucial feature in the basic organization of the double helix. For the seven years since Watson and Crick had proposed the double-helical arrangement of DNA chains, the question of whether the two chains run in the same direction or in the opposite direction had remained unanswered. The x-ray analyses

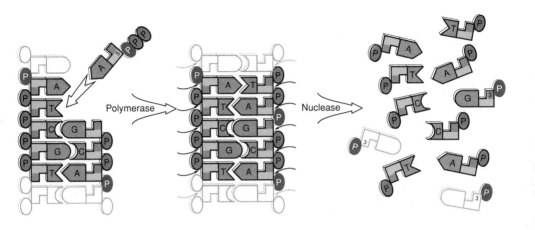

FIGURE 5-15
A procedure for determining "nearest-neighbor" sequences. DNA is synthesized from four substrates, one of which (APPP) is radioactively labeled in the innermost phosphate. Then a nuclease (DNase) is used to split the backbone of the synthetic DNA, leaving the radioactive phosphate attached to carbon 3 of the deoxyribose of its nearest neighbor. (See the text.)

upon which their model was based did not permit a choice between these two alternative orientations of the backbones of DNA chains (Fig. 5-16). When we arranged the sixteen nearest-neighbor sequences of the synthetic DNA as they might have been aligned in a pair of chains in the synthetic double helix, the result was striking. When the sequences were placed in chains oriented in opposite directions, the percentage abundances of each nearest-neighbor pair in the two chains matched almost perfectly. By contrast, in chains oriented in the same direction, the abundances of the sequences facing each other in the chains did not match at all. This relationship held for a variety of animal and bacterial DNAs we tested and left no doubt that the enzyme, in its synthetic activity, had taught us a major truth about DNA: The two chains of the DNA double helix run in opposite directions!

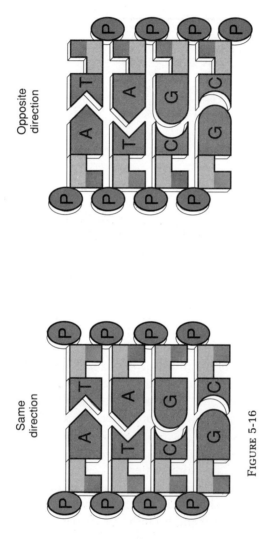

Same
direction

Opposite
direction

FIGURE 5-16
The two strands of the double helix are shown oriented in the same direction on the left and in opposite directions on the right.

Among the many sequence frequencies we determined, one stood out and has remained a baffling curiosity: the very low frequency with which G is added to C in the DNA of animals; it is only about a fourth as common as in bacterial DNAs or as adding a C to G. In view of the universality of DNA language throughout nature, this feature, unique to animal species, cries for an explanation. Does the low frequency mark places in DNA for special attention in the cell's use of a gene, or is its paucity in animal DNA simply an accident of evolution?

Bizarre Replication without an Apparent Template

Once the polymerase was pure enough for synthesis of DNA in large amounts, we wanted desperately to show, from 1957 on, that the DNA we made had genetic activity. The template we used was DNA from the bacterium *Hemophilus influenzae*, which had shown demonstrable transforming activity (capacity to transfer genes from one strain to another). Instead of an increase in this activity, we discovered a steady loss with the progress of time. Yet the enzyme preparation appeared to be relatively free of the nuclease enzymes that degrade DNA into small pieces. In trying to explain this failure, we made a startling discovery that provided an insight into the evolution of DNA and novel molecules that became a trove for many hundreds of studies of DNA.

Inasmuch as only a few breaks in a DNA chain by a nuclease might suffice to destroy its genetic activity, we needed a more sensitive method to detect them. Howard Schachman provided the device and experience for measuring the viscosity generated by polymer chains. Howard had come in 1957 from the University of California in Berkeley to spend a sabbatical year with us. He was trained in physical biochemistry and had experience in handling nucleic acids, through his earlier work with Wendell Stanley and others on the tobacco mosaic virus. One of the tools in such studies is the viscometer, an instrument that determines viscosity by noting the time it takes a solution to flow through a very narrow tube. The long thin rods of the tobacco mosaic virus and the sinuous threads of DNA make a solution viscous. A few breaks in DNA reduces its viscosity.

To find out whether our polymerase preparation might be making breaks in DNA, Bob Lehman, one day in December of 1957, placed a solution containing the enzyme and calf thymus DNA in a viscometer. To simulate replication conditions, he also included all the other necessary ingredients but one. He omitted the G substrate (deoxyguanosine triphosphate) because it had been more difficult to prepare and supplies of it were limited. The viscosity began to decrease slightly. After thirty minutes, the duration of our synthesis experiments, the viscosity had fallen 10 percent, presumably due to some nuclease splitting of the DNA.

Ordinarily, the experiment would have been terminated at this point, but, because the viscometer was in place and he had to proctor a class exam down the hall that day, Bob kept on taking readings every fifteen minutes or so. After five hours, the viscosity had fallen to half of the starting level. Then something strange began to happen. The viscosity started to rise, slowly at first and then with increasing speed. Within the next two hours, it had exceeded five times the level we started with. Clearly something had gone wrong. Perhaps specks of dirt were clogging the narrow viscometer tube. "It's surely a waste of time," one of us said, "but let's run this through once more."

When Bob repeated the experiment, he got exactly the same result. A dirty tube now seemed improbable, as did the growth of a bacterial contaminant. Hardly believing that any new DNA could be made in the absence of G, an essential building block, Bob as a further measure omitted the C building block as well; he also used a fresh batch of enzyme. Once again, the viscosity decreased gradually for four hours and then abruptly rose to six times the initial value. What next? He left out all the building blocks except for T. The result: the viscosity fell as before, but this time it remained low; there was no "Lazarus effect."

Both A and T were absolutely required, as was the enzyme. It seemed then that a polymer was really being assembled. But with calf thymus DNA as the template, how could synthesis proceed with only A and T, lacking the Cs and Gs to match the Gs and Cs as they were encountered along the template chain? "Let's leave out the DNA template." After a lag of six hours, there was a tiny increase, followed by a slightly greater rise. Within two hours the viscosity had vaulted as before to the very high level.

With radioactively labeled A or T we could show unequivocally that a new DNA was being made. After a lag of several hours, despite the absence of a DNA template, and in either the presence or absence of G and C building blocks, the enzyme built a new DNA using A and T in equal amounts. When we "seeded" a new reaction with even less than 1 percent of a previously completed one, the time lag was abolished. The product of the reaction, spontaneous or seeded, had the most remarkable structure. By nearest-neighbor analysis, we found that the As and Ts were aligned in paired chains of ten thousand or more in perfectly alternating sequence (Fig. 5-17):

$$--A-T-A-T-A-T-A-T--$$
$$\cdot \quad \cdot \quad \cdot \quad \cdot \quad \cdot \quad \cdot \quad \cdot \quad \cdot$$
$$--T-A-T-A-T-A-T-A--$$

Apparently, tiny amounts of this bizarre DNA were generated independent of any preformed template. Once formed, this novel DNA then served as a primer and template to reproduce itself. Synthesis proceeded more and more rapidly as more template was made until the supply of A or T was exhausted.

These findings immediately solved a problem that had been annoying us for months. In a number of experiments designed to test the template function of DNA, the A and T content of the product had exceeded that of the starting DNA. Now it was clear that these anomalous reactions had been run longer than usual and that the excess A and T in the product could be accounted for as these novel chains with A and T in alternating linkage. It was a relief to be able to account for discrepancies in our data, but the question we now faced was more difficult: how to explain the spontaneous generation of a DNA molecule.

In the extensive publication which appeared in 1960, three years after our discovery of the A-T polymer, we, like others, did not describe how we stumbled on it. Nor did we speculate much about the mechanism of its origin. George Beadle, a renowned biologist, and his wife, Muriel, a science writer, wrote me that they were intrigued by the spontaneous generation of a DNA molecule and impressed by the boldness of the experiments that led to its discovery. I could easily disenchant them about the "boldness," but was at a loss to enlighten them about the origin and evolutionary significance of the novel polymer.

We had a few clues. One was that a very tiny amount of the polymer was effective in priming rapid and extensive synthesis. We could measure synthesis in the spectrophotometer, because assembly of the building blocks of DNA into a double helix produces a 30 percent decrease in ultraviolet light absorption. But it puzzled us that with successive experiments the lag time became shorter and shorter, decreasing from many hours down to a few minutes. The mystery was finally solved when we paid closer attention to how we cleaned the glass reaction tubes. Our practice for years had been to rinse them with detergent thoroughly between experiments. This was not enough. Only after the action of a strong acid cleaning solution could we regularly observe the many-hour lag time for polymer synthesis. Evidently, traces of the polymer which stuck to the glass walls, though undetectable in the spectrophotometer, were sufficient to prime replication with no delay.

As an aside, this "infectiousness" of A-T polymer in our reactions made me wonder on occasion about its possible toxicity. In 1960, more than a decade before the wildly exaggerated anxieties over recombinant DNA and a genetically engineered "Andromeda strain," I thought of the consequences if the polymer somehow made its way into a cell. With access to the replication machinery in the nucleus, routine for the DNA of viruses, A-T polymer by its superior template function might usurp the place of the host DNA and be synthesized in prodigious amounts. The polymer might then be transcribed into novel messages and translated into proteins with a myriad of unpredictable effects. Breakdown of one cell would spread the polymer to others with wildfire speed. Against this scenario, I realized that cellular barriers to DNA and the ubiquitous and abundant nucleases whose function in the cell was to degrade foreign DNA would make the initial entry of the polymer exceedingly unlikely. Furthermore, infection with the polymer would not be contagious from one person to another. I decided that on a list of things to worry about, this hazard could be ranked with sunstroke in an Arctic winter.

As early as we could detect the development of the A-T polymer, it was already many thousands of units long. How did it reach this size? How small a segment would serve as a template and primer for the enzyme to make these huge chains? In 1963 Gobind Khorana, then at the University of Wisconsin, helped us solve this problem. He brought an assortment of small A-T chains that he had made by chemical synthesis. The chains contained alternating A and T in a series ranging in length from 1 A-T to 7 (A-T-A-T-A-T-A-T-A-T-A-T-A-T). (In contrast with long chains, called *polymers*, these are called *oligomers*, meaning a few, after the Greek *oligos*.) An oligomer of 7 A-T doublets can align itself alongside another to form a duplex (see below) at 37°C (body temperature). Although the chemical link between units making up a single chain is very strong and resists boiling or even higher temperatures, the force that holds an A in one chain to T in the opposite chain (a hydrogen bond) is very weak and becomes effective only when multiplied by many such bonds in succession. In the case of chains with 10 or more A-T doublets, the duplex is stable in solution up to 60°C; above this temperature, the duplex melts ("unzippers") into separate chains. For oligomers with 6 or fewer A-Ts, pairing to form a duplex is tenuous and is not readily demonstrable even at 37°C, the usual temperature for our enzyme reactions.

Would these A-T oligomers prime the enzyme to make the giant polymers, and, if so, would the temperature of the reaction be a crucial factor? The results were impressive and far more revealing about the spontaneous origin of A-T polymer than we had ever anticipated. The usual lag of 6 hours before the start of polymer synthesis which we observed when no primer was provided was not affected by adding the oligomers with only 2 or 3 A-T doublets. However, the lag was reduced to 2.6 hours by the 4-unit oligomer, and to 1.4 hours by the 5-unit; synthesis was virtually immediate with the larger oligomers, as with the polymer. When we lowered the temperature of the reaction from 37°C, the effects were profound. There was an optimal temperature for each oligomer in the series and a direct correlation with its size. For example, the optimal temperature was 10°C for the 4-unit oligomer, 20°C for the 5-unit, 37°C for the 6-unit, and 45°C for the 7-unit.

We favored a model, called *reiterative replication*, to explain these and other findings. For example, a pair of 4-unit oligomers in complete register might come apart and then come together again, but the second time only 3 units might be paired. This "slippage" by one notch generates overlapping ends, which when filled by polymerase replication creates a 5-unit oligomer:

ATATATAT *Slippage:* ATATATAT   *Replication:* ATATATATAT
TATATATA $\longrightarrow$ TATATATA $\longrightarrow$ TATATATATA

Successive rounds of slippage and replication could generate longer oligomers up to polymer size.

Why is a temperature of 10°C optimal for priming by the 4-unit oligomer? It is the best compromise between a higher temperature, which would be less favorable for formation of duplexes, and a lower one, which would freeze slippage. Exactly how one chain slips past the other or might fold back on itself and how the enzyme may affect this slippage are still uncertain. Yet the model, in essence, was a plausible way of accounting for the generation of simple DNAs, like the A-T polymer.

An overriding question about the spontaneous synthesis of the polymer was the source of the A-T oligomers to prime the reaction. We learned something from the effect of natural DNAs in hastening the onset of polymer synthesis. While most DNA samples had no detectable influence, those with a relatively high content of A and T (over 60 percent) reduced the lag from 6 hours to 1 or 2. The most remarkable example was the DNA from some species of crabs, which removed the lag almost completely. Indeed, a nearest-neighbor sequence analysis showed a large fraction (about 30 percent) of the crab DNA to be made up of stretches of A and T in nearly perfect alternation, just like the A-T polymer, except for a peppering of G and C. During evolution of some crab species, reiterative replication of an A-T oligomer within their DNA must have generated these huge tracts of nearly perfect A-T polymer.

We could now offer an explanation for the apparently spontaneous synthesis of an A-T polymer by the DNA polymerase of *E. coli*, based on what we knew about the properties of the enzyme. Generation of an A-T oligomer, starting with a single nucleotide (A or T) and adding alternate nucleotides in succession, simply by chance, was exceedingly improbable. Instead, fragments of *E. coli* DNA, persisting as impurities stuck to a very few of the DNA polymerase molecules, could serve as primers to produce the oligomers, which could then undergo reiterative replication. Undetectable by our analyses of the purified enzyme, a stretch of ATATAT within a piece of DNA attached to only one of twenty thousand polymerase molecules (among the hundred trillion present in a standard reaction) would be sufficient to get the synthesis started.

It had been troubling to us at first that reiteration of a simple sequence within a piece of DNA produced only the alternating A-T polymer and not an analogous polymer made up of G and C. Why had we not encountered the generation of the even simpler polymers in which a chain of A is matched by a chain of T or one in which a chain of G is matched by a chain of C? These concerns vanished when we did observe the "spontaneous" synthesis of each of these polymers upon varying the conditions for the polymerase reaction (acidity, presence of phosphate, attachment of certain drugs to the DNA).

Does reiterative replication as we observe it in the test tube offer clues to the evolutionary origins of DNA? More than twenty years had elapsed since I had done any experiments or thought seriously about this question. How strange that while writing these very pages of this memoir, I would receive a phone call from someone working on this subject. It was one of those coincidences that gives parapsychology its tenuous existence. The caller was completing his graduate thesis at the University of California at Irvine on the evolutionary role of reiterative replication in generating certain tracts of DNA that are virtually identical in the chromosomes of humans, mice, snakes, and flies. Despite the vast amount of attention heaped on DNA in recent years, the functions of these large tracts of monotonous DNA still remain utterly mysterious and fascinating.

FIGURE 6-1
Family photo (Stockholm, December 1959): Roger, Ken, Sylvy, myself, and Tom.

Chapter 6

Creating Life
in the Test Tube

Among the years of a life, a few stand out. To them, we relate the others by addition or subtraction. Such was 1959 for me. In May of that year, I made a month-long tour of the Soviet Union. In June we moved from St. Louis to begin our lives in California and to start the Biochemistry Department at Stanford; we were the first to occupy space in the new Medical School on the campus. In October came the announcement of the Nobel Prize in Physiology or Medicine, which I shared with Severo Ochoa. The week of events in Stockholm in December, coupled with stops in Europe before and after, was for my family and me the best party of our lives (Fig. 6-1). There was sadness in that year, too, from the loss of our fathers, Sylvy's and mine.

We knew something was brewing the day before the prize was announced. I was in Bethesda where I had given an NIH lecture. My son Roger was with me on that trip, and we were staying with our old friends Celia and Herb Tabor at their house on the NIH grounds. They had received phone calls from reporters inquiring about me. There were also reporters at the San Francisco airport when we returned home that evening. I had heard that leaks were made to the press about contenders and really gave these inquiries no serious attention. Then about five o'clock the next morning, Sylvy and I were awakened by a call from some newspaper in the East asking for my reaction to being awarded the prize. I felt elated, of course, and surprised, but not shocked.

The citations for the joint award were for Ochoa's discovery of the enzymatic synthesis of RNA and for my discovery of the enzymatic synthesis of DNA. The two discoveries occurred independently of each other, and the circumstances surrounding them were very dissimilar. In 1955, in the course of his biochemical exploration of plant and animal metabolism, Ochoa accidentally uncovered a reaction that made RNA-like polymers. In my case, as early as 1950 I had set out deliberately to find the enzymes that synthesize the nucleic acids. But I had veered in that direction after my training in Ochoa's laboratory in 1946, and so perhaps in some sense the two discoveries were not unrelated after all.

Alfred Nobel's will directs that the prize be awarded for a major discovery. Unlike election to the Baseball Hall of Fame, the Nobel Prize is not given for a lifetime record of stellar achievements, such as those of Oswald Avery, David Keilin, Stephen Kuffler, Michael Heidelberger, Harland Wood, H. A. Barker, and Charles Yanofsky, to name a few. Not only has this selection criterion eliminated many deserving scientists, but it has also made laureates of a few caught in the spotlight of a timely discovery, who did little other work of distinction before or after that for which the award was made.

The Nobel Prize has not altered my life, nor did it totally disrupt my routine even the very first day. Of course Sylvy and I were alerted for the big champagne celebration given that afternoon by Leah and Henry Kaplan, which many university notables and friends attended. But we were among the last to arrive because we had to collect one of the children from his music lesson after school.

Biochemistry, a Family Affair

Sylvy was quoted in a newspaper the next day as having quipped: "I was robbed." In fact, she had contributed significantly to the science surrounding the discovery of DNA polymerase. I cannot imagine how this work could have gone forward without her unwavering support in the laboratory and at home.

Sylvy's taste for serious science developed much earlier than mine. She was one of a very few in her class at the University of Rochester to commute from the women's Prince Street campus across town to the River campus for advanced courses in biology and chemistry. She also served as editor of *The Tower Times*, the campus newspaper. Her graduate work was in the Biochemistry Department at the Medical School, where the main focus under Walter R. Bloor was on body lipids. She was among the earliest to use radioactive phosphate to trace the metabolism of phospholipids but was discouraged by the lack of local interest in applying this technique to enzymes and intermediary metabolism. Her first job was at the National Cancer Institute, where she began work in 1942 with Jonathan Hartwell, an organic chemist, and Murray Shear, a biologist, on the synthesis of novel carcinogens and their effects on mice.

We first met in Rochester, became better acquainted in Bethesda, and were married in November of 1943. When our three boys were small, Sylvy edited books at home for Interscience Publishers (now part of Wiley), among them *Advances in Enzymology* and George Hevesy's *Radioactive Indicators*. By the time we moved to St. Louis in 1953, Roger was six, Tom was five, and Ken was three—old enough, we thought, for Sylvy to resume full-time research in my laboratory. She discovered the enzyme that makes the huge polyphosphate chains that serve as phosphate and energy reservoirs; the wide significance of this neglected area of energy metabolism is just now being appreciated. Sylvy thought about experiments and loved to do them. Her high intelligence led to key insights at many stages of our work on replication, including the metabolic modifications viruses introduce to make their unique DNAs.

Our children were often with us in the lab after school let out and on weekends. It was a busy, congenial atmosphere in which postdoctoral fellows (Paul Berg, Bob Lehman, Maurice Bessman, Julius Adler) and others gave them fond attention. When asked at age nine what he wanted for Christmas, Roger said: "A week in the lab."

Almost always, I took one of my boys with me on lecture and meeting trips. Before grade school, there were two-day visits to New York, Washington, or Boston (where we had friends and relations), with a half-day added for sightseeing. When they were older, weeklong trips to Yale, to the Waksman Research Institute at Rutgers, or to the Medical College of Georgia in Augusta included attendance at school with a friend's child. The boys were wonderful companions on a dozen such trips, each delicious and memorable.

Once, after attending a Welch Chemistry Conference in Houston with Roger, we went on to see New Orleans. We arrived late at night at our hotel and found the late Feodor Lynen (the famous and irrepressible German biochemist) in the lobby, lacking a room. I invited him to share ours. When we started to undress for bed, Lynen said: "It's only midnight, we must see New Orleans." "But Roger is only thirteen," I pleaded. The vote was two to one, and off we all went to the jazz spots on Bourbon Street, not to return until 4 A.M.

Roger had a princely preparation for biochemistry, spending successive summers after high school and during college doing nucleic acid enzymology with Paul Berg, bacterial genetics with Charles Yanofsky, and organic chemistry with Carl Djerassi. When Charles Richardson (later chairman of the Biochemistry Department at Harvard Medical School) came to Stanford as a postdoctoral fellow in 1961, Roger, then in high school, taught him how to purify DNA polymerase but was embarrassed at not knowing what ammonium sulfate was. The last free summer of high school he spent on the science schooner *Te Vega*, collecting Pacific fauna.

Not having to devote college time at Harvard to learning about research, Roger could take the advanced courses in chemistry, physics, and mathematics that I never had and always missed. When he returned from Harvard to do graduate work at Stanford, he lived at home, where lab problems and progress were common dinner topics. In applying the magnetic resonance techniques pioneered by his graduate advisor Harden McConnell to biological membranes, Roger introduced methods that led to a profound insight.

Membranes are made up of two palisade-like layers of phospholip-

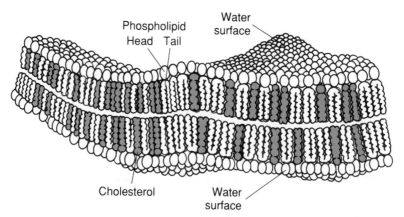

FIGURE 6-2
The cell membrane, composed of phospholipid molecules, is organized as a pair of layers with interspersed cholesterol molecules (colored) and protein molecules (not shown). In each layer the fatty part of the phospholipid faces away from the water surfaces on the inside and outside of the cell.

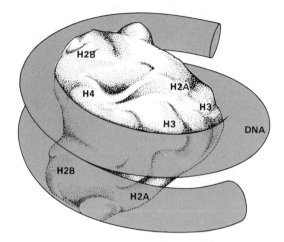

FIGURE 6-3
DNA in chromosomes is wound about a core of proteins (histones), an octet of two each of Histones 1, 2, 3, and 4. A succession of these nucleosome units resembles beads on a string.

ids (Fig. 6-2). The phosphate end of each phospholipid molecule in each array faces the watery worlds of the inside or outside of the cell. The long fatty acids of each phospholipid, with cholesterol often interspersed, mingle freely in the fatty world of the middle of the membrane. Roger discovered that individual phospholipid molecules move about freely, exchanging places with a neighbor ten million times a second; in contrast, the "flip flop" of a molecule from one layer to the other happens rarely, perhaps once in many days. These facts are crucial to understanding the fluidity, rigidity, and functions of the cell membrane.

For further training in structural chemistry, Roger spent three years at the renowned Molecular Biology Laboratory in Cambridge, England, working on a long-standing problem: How are DNA and associated proteins (histones) organized in the chromosome? By using gentle biochemical methods, he found that the histones form spherical octets with the DNA wrapped around them (Fig. 6-3). These fundamental units, called *nucleosomes*, form a chain of beads which is further folded in the compact chromosome. At Stanford since 1978 (and now chairman of the Department of Cell Biology), Roger has directed his research to the problem of how the nucleosome and other molecular arrangements govern the expression of genetic information.

Tom's talent for music and devotion to the cello from the age of eight precluded any special attention to science. At sixteen, he was selected by the late Leonard Rose, a premier cellist and teacher, for the Juilliard School. At my urging, Tom also enrolled as a full-time student at Columbia College and took courses in biology, chemistry, and physics as part of a liberal arts curriculum. While the deans at both schools assured me of their enthusiasm for this combined program, Juilliard insisted he fulfill the science requirement with their evening course in psychology, and Columbia gave him no credit for humanities, despite four years of Juilliard courses in music theory, composition, and performance.

When the occupational trauma to his left index finger produced painful neuromata (nerve-end tumors) and forced him out of musical performance, Tom decided to try laboratory work. No novice ever made a more sensational start, as I will describe in Chapter 7. He then did graduate work in biochemistry at Columbia and postdoctoral research, first at Princeton and later at the Molecular Biology Laboratory at Cambridge, where, with the guidance of Peter Lawrence, he took up *Drosophila* genetics and development.

Tom has a great gift for experimental work and a sense of purpose that has made him a leading figure among those seeking the molecular basis of developmental biology (that is, embryology). He is now a professor in the Department of Biochemistry and Biophysics at the University of California in San Francisco.

Ken's early exposure to biochemistry was evident when at the age of eight, in a discussion of what acid to use to remove cement spots, he brightly suggested the only one he knew: "Nucleic acid." Ken worked with me one summer on bacterial sporulation and then for six months (after finishing high school early) with Arturo Falaschi at the Molecular Biology Laboratories of the University of Pavia in Italy. But science was not to be his profession. When exposure to archeology in two summer digs in Israel left him unhappy with the fragmentary and inconclusive basis of this discipline, he turned to architecture, where he found a means of expressing his taste for design with less ambiguous results.

Ken did not escape science entirely. As an architect, he specializes in the design of laboratories and buildings for research in biomedical science and biotechnology. Most architects engaged in planning a laboratory building are intimidated by the arcane activities of their clients. As a result, they lavish their attention on the "dry spaces"— the offices, libraries, and conference rooms—where they believe the truly creative work is done. They leave to engineers and lab furniture suppliers the outfitting of the "wet spaces"—the drab laboratory cells, unchanged in fifty years, where the technicians test the ideas spawned by the scientists.

Ken knows that molecular biology is not a theoretical science and that the advances his family and friends have made are the result of spending almost all their waking hours in those unattractive wet spaces. Why not make them bright, open, sunlit, and colorful? Why not make the laboratories inviting places to work, mingle, and feel as one would in a comfortable house? Such designs can be even more economical in the use of space and equipment, while still being "scientist-friendly."

Sylvy and I had complete trust in our children and confidence in their wisdom and abilities. They shared access to our bank account from the time they entered college and reciprocated our feelings toward them. Perhaps this shared sense of trust helped them to cope with the turbulence of the 60s and 70s without turning to drugs or dropping out to "search for identity." Beyond the deepening friendship that came with their maturity, what has given me even more comfort is their intense loyalty to one another.

The Stanford Biochemistry Department, an Extended Family

I have been fortunate in my family life—at home and in the laboratory, too. The Biochemistry Department at Stanford seems strange to most people—a small faculty, together for over thirty years, sharing resources communally and focusing on a single subject, DNA. Even those familiar with the department believe it started with this novel design and lived happily thereafter. Not really. There was no such plan or goal. More people have left than stayed. Those who stayed evolved the patterns and interests that gave this academic family its unique shape and character.

The Stanford Biochemistry Department began to coalesce in 1953 while I was still in St. Louis at the Washington University School of Medicine. Flattered to be chosen as professor and chairman (at age 34) in a school which was then the most prestigious in medical research, I left Bethesda for St. Louis on January 21, Inauguration Day, quipping that with President Eisenhower's arrival, Washington

FIGURE 6-4

The Microbiology Department of Washington University School of Medicine in St. Louis (circa 1956). *1st Row:* J. McLeary, K. Horibata, M. Walsh, J. Hurwitz, A. Kornberg, E. Battley, D. Hogness. *2nd Row:* I. R. Lehman, M. Bartsch, A. D. Kaiser, S. Johnson, M. Cohn, P. Berg, E. Simms, J. Ofengand, G. Bugg. *3rd Row:* H. Wiesmeyer, O. Ward, E. Holmes, D. Daniels, H. Morales, L. McKeown, M. Bessman, E. Stonehill, V. Johnson.

would be too small for both of us. My first appointments to the new Department of Microbiology were my two postdoctoral fellows at NIH—Osamu Hayaishi, who came as an assistant professor, and Irving Lieberman, an instructor. Paul Berg arrived that fall as the first postdoctoral fellow. After graduate work in biochemistry at Western Reserve University in Cleveland, he had been strongly advised by Harland Wood, the chairman of that department, to spend a second fellowship year in Carl Cori's laboratory in St. Louis after the first with Herman Kalckar in Copenhagen. Paul rebelled against enduring the St. Louis climate and chose instead to work with me at NIH. Imagine his reaction when I wrote him in Copenhagen that I would be moving to St. Louis the next year. How fortunate for all of us that this time he decided to come to the "Gateway to the West."

The laboratories in the Microbiology Department were fifty years old and decrepit. Bare light bulbs hung from twenty-foot-high ceilings, electric outlets were sparse, the tiny sinks leaked, and the wooden bench tops were pitted and corrugated. There were many keys, keys for every room and cabinet. I threw them all away. But the labs! Why had I ever left the bright, new, well-equipped labs at NIH? Ollie Lowry, a gifted experimentalist, genial colleague, and chairman of the Pharmacology Department, tried to console me: "My department was in as bad shape, but with some wiring and fixing I did on weekends, it turned out fine." "Ollie," I moaned, "I can't wire a lamp, let alone a department." Three laboratories had been tardily refurbished, but one needed for Paul Berg was still lacking. Finally, I told the Dean that if it was not ready in a month, I would leave, and Carl Cori told him I meant it.

In the next three years, seven people joined the faculty (Fig. 6-4). Melvin Cohn, David Hogness, and Dale Kaiser came after their fellowship training with Jacques Monod and François Jacob at the Pasteur Institute, then the center of bacterial genetics. Later, Robert De Mars came from Salvador Luria's group at the University of Illinois, and Jerard Hurwitz from the NIH. Paul Berg and Robert Lehman remained after completing their fellowships with me. Our faculty had training and skills in immunology, virology, and microbial metabolism, but we lacked a parasitologist and tried unsuccessfully to find one with a biochemical orientation. During this interval, four people left the faculty: Hayaishi to return to Japan via the NIH., Lieberman to the University of Pittsburgh, De Mars for army service at Walter Reed Hospital, and Hurwitz to New York University Medical School.

In these embryonic years, the key cultural features of the department were developed. We shared our limited funds, initially supplied as grants to me and then to the others as their research programs matured; all decisions for significant expenditures and space allocations were made communally. We shared our scientific interests in daily noon seminars; each of us felt impelled to understand every topic and accepted the challenge to present material in one another's special areas. Discussions were sharp and uninhibited but always friendly. We knew the status of one another's experiments well enough to understand "It worked" as a passing remark. We shared eagerly in the teaching, involving our postdoctoral fellows as well.

We taught microbiology to medical students in their second year and emphasized the biochemistry and genetics of disease organisms rather than their diagnostic features. The students rebelled at this radical change and, abetted by two faculty members from the previous department, called our course, Biochemistry II, an extension of the course endured in the first year. For lack of a supportive textbook or precedents in other schools, this student resistance lasted most of our six years in St. Louis. Yet after we left, Herman Eisen and his new faculty kept our curriculum intact. With the appearance of *Microbiology*, an impressive, modern textbook (by Davis, Dulbecco, Eisen, Ginsberg, and Wood), the battle for basic microbiology was won. Many students we taught in those years later told us how grateful they were for a firm foundation in microbial disease, having easily acquired diagnostic details in their subsequent clinical training.

In April 1957 Henry Kaplan, a friend and Chairman of the Radiology Department at Stanford, mentioned Stanford's interest in me for the chairmanship of its new Biochemistry Department. When two months passed without hearing anything further, I felt relieved that my settled life in St. Louis would not be challenged. Then Sylvy and I were invited to visit Stanford. We tried to look beyond the stunning reception—beyond the flattering fact, for example, that when we stopped for gas at Rickey's Motel on El Camino Real, where we were housed in an elegant suite, the attendant responded: "Yes, sir. By the way, Dr. Kornberg, I enjoyed your seminar very much." On returning to St. Louis, I described the Stanford prospects to my colleagues, and I had to admit that the case was compelling.

The Stanford Medical School would be moving in two years from San Francisco to the campus in Palo Alto. In the handsome new buildings designed by Edward Durell Stone, architect of the U.S. Embassy in New Delhi, the Biochemistry Department was assigned a spacious floor which I could design. This unpopulated department would be staffed with colleagues of my choosing. Now a legitimate biochemist, I could teach and practice my subject without disguise and appoint the physical and organic biochemists whom I very much missed. No longer would I be the muted, junior member of the tradition-bound Executive Committee of Washington University Medical School. Rather, I would have a large share in policy and faculty selection in the renaissance of science and medicine at Stanford promised by J. E. Wallace Sterling, the President, and Frederick E. Terman, the Provost. Having escaped to California from St. Louis for several summers, Sylvy and I were keen to live permanently in that most agreeable climate and geographical setting, one which would also be attractive to our children, future students, and colleagues. Instead of remaining in the "Gateway to the West," we might now be housed in the manor itself.

The enthusiasm of all my departmental associates to join me ultimately made my decision easy. Robert (Buzz) Baldwin, twenty-eight years old and already recognized for contributions to the physical chemistry of proteins, agreed to come from Wisconsin. This brought to seven our starting faculty group of Berg, Cohn, Hogness, Kaiser, Lehman, and myself. Our impending move also persuaded Joshua Lederberg, sought by everybody, to leave the University of Wisconsin to start a Department of Genetics in the Stanford Medical School. (His preference then, and in subsequent years, was to be a member of the Biochemistry Department, but I felt that his leadership in genetics deserved a larger stage.) Charles Yanofsky, at Western Reserve University, upon learning of our plans, decided to accept an offer from the Department of Biological Sciences. I was also drawn into the recruitment of new faculty for the Chemistry Department, which, with the selection of William Johnson as chairman and then Carl Djerassi, Paul Flory, Henry Taube, Harden McConnell, and Eugene van Tamelen, vaulted into national prominence.

I dreaded telling Carl and Gerty Cori, who had been responsible for my coming to St. Louis, that I had decided to leave. Carl was angry and sputtered: "Where will you go on vacation?" Gerty quickly intervened: "Carlie, perhaps we too should have gone to California when we had the chance." I was also sorry to leave Ernie Simms, who had become a devoted and effective research assistant. His roots in St. Louis were too deep for transplanting. Disadvantaged

FIGURE 6-5
The Biochemistry Department of Stanford University School of Medicine (circa 1966). *1st Row:* J. Kriss, D. Hogness, L. Soll, I. Scheffler, P. McPhie, D. Nelson, E. DuPraw. *2nd Row:* B. Olivera, M. Deutscher, K. Gray, G. Stark, E. Sherberg, D. Kaiser, A. Kornberg, R. Lehman, L. Stryer, R. Baldwin, P. Berg, M. Goulian. *3rd Row:* P. Chambon, B. Egan, R. Sternglanz, P. Primakoff, P. Bayley, W. Galley, W. Folk, C. Scandella, Z. Hall, K. Collins, N. Cozzarelli, J. Finsterbusch, J. Scheufler, L. Bertsch, H. Epstein, G. Hobom, L. Foster, C. Huebner. *4th Row:* D. Loskutoff, J. Champoux, E. Padilla, T. Mackinlay, P. Englund, T. Jovin, M. Pearson, R. Freedman, R. Doherty, F. Welland, J. Foulds, C. Hill, B. English, T. Broker, L. Cromwell.

by lack of college training due to his black origins, he more than compensated by his intelligence, motivation, and good judgment. Washington University later awarded him an associate professorship with tenure.

The two lame-duck years in St. Louis were awkward. There were some resentments among the faculty despite our having brightened and equipped the place, largely with external grants, and having provided a clean slate for appointment of a new faculty. In June 1959, twenty-two people—including Esther Sherberg, my future administrative secretary, Hilbert Morales, our laboratory manager, Peter Hoefer, sculptor and instrument maker (later, founder of the highly successful Hoefer Scientific Instruments), and five children— set out for the promised land.

Among decisive events—"crises" in the autobiographies of politicians—I will select three in the history of the Biochemistry Department at Stanford. One was in 1961 when Mel Cohn had decided to become a founding member of the Salk Institute in La Jolla and we sought to replace him with a biochemist skilled in mechanisms of enzyme action. After interviewing several candidates, we found someone very attractive to three of us, myself included, but less so to the other three members of the faculty. Without deliberating, I proposed that we keep looking until we found someone who appealed to us all. I had chosen the faculty until then, and although I still had this authority, I no longer felt comfortable with it. On reflection, the importance of this action was to make it clear to each member of the faculty that he had an equal voice in choosing a colleague, the most important decision a department can make. George Stark and Lubert Stryer soon came along. Both seemed right to all of us, and so we engaged them both (Fig. 6-5).

Another crisis came in 1969. We needed more space to accommodate the growth of each of the faculty research groups and were granted several rooms in a contiguous wing. At this juncture, Lubert Stryer wanted this new space assigned to him—a reasonable share of the total available. Doing so, however, would have disrupted our established practice of sharing all the departmental space. Our custom of mixing students and fellows from several groups in the four-person laboratories has without exception been regarded by them as one of the most rewarding features of their stay at Stanford. Despite our wish to adjust to individual faculty styles, this issue was so crucial to the rest of us that we reluctantly let Stryer accept an offer from Yale that fulfilled his wishes for more autonomy.

A third decisive event also took place in 1969. That year I resigned as chairman of the department, having served as chairman at Stanford for ten years and before that for six years in St. Louis. Chairmen of medical school departments were expected to hold office until retirement, but with the broadening of the pyramidal structure of departments, I felt such a life sentence to be unrewarding for both the individual and the institution. We had all shared in the direction and management of the Biochemistry Department, and several members were highly qualified to assume that responsibility. Paul Berg was the first choice; he felt comfortable in the role and showed initiative in inaugurating an annual research "retreat" at Asilomar, which became the model for departments nationwide. After five years, the chairmanship rotated to Bob Lehman, then to Dale Kaiser, and is now held by David Hogness. At no time since I stepped aside has the style of the department changed, nor have I felt less involved in the operations and policies that define its style.

Our responses to noncrisis problems have been at least as important as the way we have managed decisive events. One recurrent issue has been size. Despite many pressures and temptations, we have kept the faculty small, at about half the number in comparable biochemistry departments. Rather than trying to cover two dozen subject areas, we concentrated on essentially one: nucleic acids and the proteins they interact with. Yet, I felt our scope was broad because we embraced a spectrum of disciplines, from physical chemistry to genetics. To cover other areas of biochemistry, we helped in the recruitment of biochemists for leadership roles in other basic science and clinical departments of the university.

Even though the size of the faculty remained small—increasing to nine with the arrival of Ron Davis in 1972 and Doug Brutlag two years later—each of the faculty research groups expanded greatly, reaching an average size of ten; a bulge beyond that size in one group was generally balanced by contraction in another. A typical group has five or six postdoctoral fellows, three or four graduate students, a senior research associate or more generally a research assistant. As research associates, LeRoy Bertsch in my group and Marianne Dieckmann in Paul Berg's provided the continuity needed to balance the constant turnover of students and fellows. Bertsch has given my group patient, wise, and informed guidance for over 25 years, as has Dieckmann to the Berg group. Expansion of the department required addition of secretarial, shop, and custodial people and, most important, a highly competent business manager who functioned, in effect, as a deputy chairman. Our annual departmental photograph showed 30 faces in 1959, 75 in 1969, and 100 in 1979. The census in 1988 was 122, and the annual budget was in excess of ten million dollars.

Some precious things were lost in this otherwise healthy growth. One which I fondly recall was the monthly evening research seminar, with everybody crowded into the living room of my house. After several years we had to move the meeting to the larger library space in the department. An essential ingredient escaped, and the seminars quickly vanished. The noon journal-club seminars (reviews of published reports from other laboratories) survived longer but were held only twice a week rather than daily as in St. Louis. Eventually, these were also squeezed out by the pressure of competing seminars from visitors to the many related departments in and near Stanford. To remain cohesive and aware of each other's ongoing research, the faculty meet around a small table at lunch on Wednesdays, but the struggle against the commitments of busy schedules never eases.

A recurrent problem has been overspending our budget. Periodically, we have had to declare a moratorium on buying equipment and expensive reagents and to appeal for initiatives in finding additional grant support. Without strict accounting, it was natural for a faculty member to believe that he was not overspending. Nevertheless, we resisted the temptation to adopt the conventional "every tub on its own bottom" practice. The more affluent scientist recalled being helped when he was getting started, and we all agreed that the pursuit of a research direction should not depend on its fashionability for funding. The communal sharing had proven over and over again its effectiveness in promoting science and good fellowship, economies of funds and resources, and ability to weather occasional concerns about inequities.

Perhaps the most serious and persistent concern in recent years has been aging of the faculty and how to prepare for the future. I never thought it useful to plan even five years ahead, but still the calendar had to be noticed. On the 25th anniversary of the department in 1984, six of the faculty of ten were 57 years or older. Remarkably, their productivity during the previous decade had been maintained, they were near the forefront of new directions, and they continued to attract able students and postdoctoral fellows. The enthusiasm to go on is the prime ingredient for scientific vigor, and this is fueled by examples of the colleagues around you. In the meantime, younger people have been added to the faculty so that at the turn of this century, when the starting team will have been officially retired, the new group in place will have the same chance for devotion to science and the pleasure of sharing it with a collegial family.

Spores: An Extreme Life Form

Those familiar with my research career are aware of my intense concentration on a single subject—the enzymatic synthesis of DNA—and the blinders I have worn to maintain this focus. Nearly forgotten now by us all are the eight years, in the midst of the DNA work (1962–70), when half of my research effort was devoted to an arcane subject, the development and germination of spores.

During my tenure as chairman of a department of microbiology (1953–59), I had become interested again in spores as agents of disease: anthrax, tetanus, botulism. I constantly remembered with deep anguish the ghastly *Clostridium perfringens* (gas gangrene) spore that killed my mother within a day of a "routine" gall bladder operation in 1939. Now I could look beyond the "bad" spores to the vast array of innocent species whose mysterious biology and biochemistry fascinated me.

One of the last-ditch arguments of the spontaneous generationists was the outgrowth of bacilli from dry hay infusions (soups) that had been boiled (at 100°C) for as long as two hours. How could one account for the disappearance and reemergence of these hay bacilli (*Bacillus subtilis*) after treatments that kill all known microbes? Ferdinand Cohn (1828–1898), a German botanist, solved the mystery by observing that in these heat-resistant cultures each bacillus had inside it a small, spherical, highly refractile (bright) body, a spore (Fig. 6-6); each spore, after boiling, could still generate a new bacillus when placed in a nutrient medium.

How could spores be killed? Temperatures of 120°C—achieved in an autoclave by steam under pressure—were needed. Another way of destroying spores was discovered by the English physicist John Tyndall (1820–1893), who by now understood the life cycle of the hay bacillus. After boiling the bacilli for 5 minutes to destroy the vegetative cells, he allowed the resistant spores to germinate for a brief period and then subjected the emergent sensitive bacilli to

Vegetative
cells

Endospores

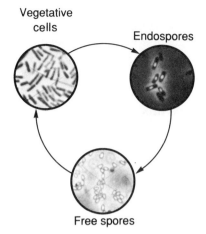

Free spores

FIGURE 6-6
The cycle of sporulation and germination. The vegetative cell, under adverse conditions, develops a spore (endospore), which is released (free spore) and remains dormant until favorable conditions signal its germination into a vegetative cell.

another boiling. Repetition of this procedure completely sterilized a medium.

What others had learned about the hay bacillus helped Koch explain a mysterious aspect of anthrax. Farmers knew that cows and sheep might remain healthy in one pasture but succumb promptly to anthrax when taken to graze in another. There were lovely green fields in mountainous Auvergne in France where no one dared allow a flock of sheep. Koch discovered that spores of anthrax bacilli from the buried carcasses of diseased animals could lie dormant in the soil for many years. These spore-ridden fields were the "occult" source of anthrax infections of animals that had grazed them down to the ground.

The menace of spore-formers was later recognized in bacilli that cause tetanus and botulism. Spores of *Clostridium tetani* in the soil, introduced through punctured skin, germinate in the body, multiply, and release the toxin that causes the fatal lockjaw disease. Spores of *Clostridium botulinum* in improperly canned food germinate and release the poison that causes violent abdominal cramps and death.

Spores as agents of disease had been studied for years, but what fascinated me most was how a spore is made and how its chemical organization endows it with such astonishing abilities. What signals a growing bacillus that hard times lie ahead and that it is prudent to enter a state of dormancy and hibernation rather than divide to form two vulnerable daughter cells? What essential cellular components are assembled, and how are they condensed and encased in layers of envelopes to produce this state of hibernation, in which the spore can withstand extremes of heat, desiccation, disinfectants, and ultraviolet rays lethal to the cell? How does a spore manage to remain utterly dormant for many years and then to revive in one minute, when exposed to a substance that signals conditions are right for growth into a new bacillus? To understand these processes in a bacterial spore was an end in itself, but in addition I believed then, and still do, that this knowledge would contribute in a major way toward understanding the embryonic development of animals and their response to environmental stresses.

In 1962 I started work on spores by examining how DNA is stored in the spore, what replication machinery, if any, is included, and then how the DNA is used when the spore is called upon to generate a new cell. My main interet in DNA replication was sustained, specifically in the enzymes used by E. coli to assemble the nucleotide building blocks into its DNA.

During the eight-year period of this work, a succession of ten postdoctoral fellows and three graduate students spent an aggregate of twenty-three years working with me on spores. We published twenty-six scientific papers describing a number of basic biochemical features of spores. For example, we found that a spore contains all the vital information and machinery needed to start a new cell, differing from the parent principally in the choice of fuel reserves, absence of extra baggage, and the way the cellular contents are arranged and encased. Yet, we were still remote from satisfactory answers to the global questions that had attracted me at the outset.

Although my fascination with spores remained, I abandoned them for several reasons. I came to realize how much more complex sporulation and germination were than I had initially assumed. Several hundred genes are devoted to these processes, and we knew very few of them and their protein products. Only the scantiest information was available about the biochemistry and physiology of spores, and hardly any one else in the world seemed to care. The little research on spores, then and even now, was largely of a practical nature— how to destroy spores in food canning or how to use them as pesticides on crops.

Beyond the discouragement and loneliness of working on a tough problem in an unfashionable area was the distraction of having the other half of my research group of eight engaged in the more glamorous and productive work on DNA replication. Because of my own ambivalence, I offered no resistance when a member of the sporulation group would defect to the replication team. Finally, after this eight-year siege, I too gave up. Eventually only two of the group, all of whom have left Stanford, continued their interest in spores.

Progress in science depends on how vigorously the field is cultivated. In contrast with sporulation, interest in cancer is enormous. Hundreds of laboratories worldwide attract many thousands of scientists, including some of the brightest, to unravel the processes responsible for malignant growth. Yet studies of sporulation are also deserving of resources and talents. Sporulation, dormancy, and germination are fundamental processes in nature, more accessible to incisive examination, and, if better understood, might yield as much information relevant to the cancer process as some of the massive programs on tumor-bearing animals.

Viruses: At the Border of Life

I recall a lecture on viruses as being the most memorable one of my medical student days. It was in 1938 and was given by George Packer Berry, then Professor of Bacteriology at the University of Rochester and later Dean of Harvard Medical School. He described the spectrum of living matter. At the lower reaches, the simplest of the bacteria adjoined the most complex of the viruses and farther down, the simplest viruses merged with large molecules (Fig. 6-7). I was thrilled by a unification of Nature without lines of demarcation or mysterious rifts. I felt liberated from the earlier need to separate animate from inanimate, animal from vegetable, cells from molecules.

Viruses are known to us today as the simplest of microbes, too small to be seen even with the thousand-fold magnification of the light microscope. Lacking the appropriate metabolic machinery, viruses, unlike bacteria or animal cells, are inert in a nutrient medium. They must enter and parasitize cells in order to reproduce. They commandeer the cell's machinery to manufacture many hundreds or thousands of new virus particles, often destroying the cell in the

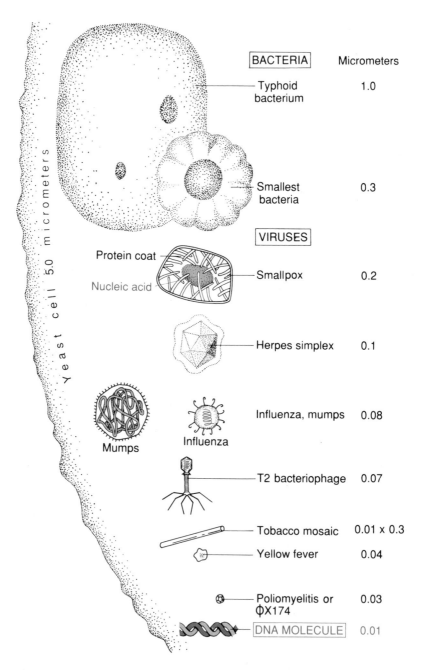

	Micrometers
BACTERIA	
Typhoid bacterium	1.0
Smallest bacteria	0.3
VIRUSES	
Smallpox	0.2
Herpes simplex	0.1
Influenza, mumps	0.08
T2 bacteriophage	0.07
Tobacco mosaic	0.01 x 0.3
Yellow fever	0.04
Poliomyelitis or φX174	0.03
DNA MOLECULE	0.01

FIGURE 6-7
The size range of molecules, viruses, and bacteria reflects their complexity and shows no significant gaps or demarcations.

process. The word virus (Latin, poison) had, a century before, encompassed the whole range of toxic agents that cause disease, bacteria included. The current, restricted use of the term developed in a curious way.

The early bacteriologists who followed the fate of pathogenic bacteria deposited in soil discovered that rain and effluents did not wash them into the lower layers. From this sieving effect of soil came the idea of using compacted earth and unglazed porcelain as filters to remove bacteria from solutions. In some instances, an agent which could transmit disease to experimental animals passed through these filters. No microbe could be seen in the microscope nor could the agent, called a "filtrable virus," be grown in culture media.

Filtrable viruses remained in limbo until three things happened: (1) they were isolated in pure form and shown to be simply a nucleic acid (RNA or DNA) enveloped in a protein coat. (2) they were actually seen in the electron microscope (with a magnifying capacity one-hundred-fold greater than the light microscope); and (3) they were cultivated inside cells grown in culture media. Soon a huge menagerie of plant and animal viruses was discovered. They had unique shapes and were of widely differing size (Fig. 6-7). The virus of poliomyelitis was seen as a multifaceted sphere (a geodesic-dome-like polyhedron), and the virus of tobacco mosaic disease as a long, thin rod. That tiny bacteria were also beset by viruses (called bacteriophages—or simply phages—because they devour their hosts) was disbelieved until they, too, were also seen in the microscope and in a great variety of forms. Jonathan Swift's prescient rhyme deserves recalling:

> So, naturalists observe, a flea
> Has smaller fleas that on him prey;
> And these have smaller still to bite 'em;
> And so proceed *ad infinitum.*

Fifteen years after I was introduced to viruses as a medical student, they again sparked my interest, this time in their DNA and its replication. Animal cells take 6 to 8 hours to reproduce their DNA; bacteria take only forty minutes, but upon viral infection (as when the T2 bacteriophage invades *E. coli*) the rate of viral DNA synthesis is ten-fold greater than the DNA of the uninfected cell. When I began to wonder how a virus might achieve this, I realized that nothing was known about how any cell synthesized its DNA. Because it is generally best to explore the biochemistry of a process in a place and at a time that it is most active, I selected T2 phage-infected E. coli as the place to look for clues to DNA synthesis. In my initial studies, the cell-free extracts of phage-infected E. coli cells proved to be even less active in DNA synthesis than those from normal cells. Yet the use of phages, as I shall describe, was to prove crucial in clarifying many features of the exceedingly complex process of chromosome replication.

Synthesis of a "Living Molecule"

"Why have you been unable to replicate the genetic activity of the DNA template?" Invariably the same question was asked after each of the many seminars I gave during the ten-year period (1957 to 1967) that we continued our studies of the DNA synthesized by the enzyme we called DNA polymerase. Occasionally I tried to forestall the embarrassing question by showing a gag slide (Fig. 6-8) in which the biological activity of DNA in a skin cream was measured by the pulchritude of a model; the assay, I explained, was still too erratic for routine use. Of course, we had tried many times to replicate the DNAs from Pneumococcus, Hemophilus, and Bacillus subtilis, each with readily measured transforming activity. These DNAs performed adequately as template-primers for net synthesis of new DNA, but there was always a net loss rather than an increase of biologic activity. The experiments were doomed to fail for reasons that I learned only years later.

The solution to our problem was to come from studying the replication of the tiniest of bacterial viruses. One of them was isolated from Parisian sewage by N. A. Boulgakov in the laboratory of M. F. d'Hérelle. He named it φX174: the Greek letter phi is for phage; X174 is the number assigned to a member of a particular strain among many Boulgakov isolated that infect and destroy typhoid bacilli. (It had been a vain hope for two decades that the bacteriophages discovered in 1917 by Twort in England and d'Hérelle in France would prove effective in combating intestinal infections by typhoid and dysentery bacteria.)

INNOXA
PARIS

CREME ACIDES NUCLEIQUES

la crema **MIRACOLO**
dall'effetto **IMMEDIATO**

La crema che RIVOLUZIONA quanto
è stato fatto finora contro L'INVEC-
CHIAMENTO del derma

2 PREMI NOBEL (1959 e 1962)
hanno decretato il SUCCESSO PIÙ
STRAORDINARIO allo studio e al-
l'applicazione degli ACIDI NUCLEICI

FIGURE 6-8
Advertisement in an Italian
journal (circa 1963).

Viral infection of a bacterial cell can be made obvious to the naked eye quite easily. A droplet of a very dilute solution of viruses is spread over the surface of agar upon which a lawn of contiguous bacteria has grown. In an instant, a virus attaches to the surface of a bacterium and injects its DNA, which takes over the cell's replicative machinery and is multiplied many times over. Within twenty minutes, the cell bursts open to release two hundred or more particles, each ready to infect a nearby bacterium. After a few hours, the many billion descendants of a single virus have destroyed a comparable number of bacteria. The death of the bacteria is marked by a neat circle of clearing in the bacterial lawn about the size of a pinhead. By counting each clear zone (which is called a *plaque* and marks the spot where a single virus first entered a bacterium), the number of viruses in the original droplet of solution can be estimated accurately.

The extremely minute size of phage ϕX174 piqued the interest of Robert Sinsheimer in the early 1960s. A biophysicist, he had moved from the Physics Department of the University of Iowa to the Biology Division of the California Institute of Technology. Imagine his surprise and delight when he discovered several unusual features of the DNA of this long-neglected creature. The DNA was exceedingly small. It contained only about 5,000 nucleotides, enough to supply information for five or so average-sized genes, compared with over a hundred genes in the larger phages and four thousand in *E. coli*. Among other bizarre physical properties was the strange composition of its building blocks: A 25%, T 33%, G 24%, and C 18%. Clearly, there was no match of A to T nor of G to C. The best and ultimately correct interpretation was that the tiny phage chromosome is a single strand rather than the double helix found in larger phages and in all bacterial, plant, and animal cells.

When we tested some phage ϕX174 DNA Sinsheimer had sent us, its behavior as a template for DNA polymerase was reassuring to us all. With the extent of synthesis limited to a small percentage of the input DNA, the composition of the product was unlike that of ϕX174 DNA but exactly what we predicted from base pairings with the template: A 33%, T 25%, G 18%, and C 24%. When synthesis was extended to a level of 600%, the composition and the frequency of nearest-neighbor sequences showed that both the single-stranded ϕX174 DNA and the complementary strand product had served as templates; each was matched by synthetic strands oriented in the opposite direction (see Chapter 5). This double-stranded DNA product directed further replication as effectively as the common DNA templates. In brief, the enzyme had converted the atypical single-stranded DNA into a conventional duplex DNA.

With more studies of ϕX174 DNA, Sinsheimer and his coworkers found three other salient facts. The DNA chain lacked ends and so it must be circular. Second, within seconds after entering *E. coli*, enzymes in the cell, much like our DNA polymerase in the test tube, converted the viral DNA from a single strand to the common double helical structure (Fig. 6-9); the content of A exactly matched that of T, and G was equal to C. Finally, the DNA could, by itself, infect a cell. Without its protein coat, which the virus uses to attach itself to the surface of the cell and which makes every particle infectious, uptake of naked viral DNA by a cell is a rare event, ten thousand times less efficient. Nevertheless, this infectiousness by free DNA could still be measured with confidence and accuracy.

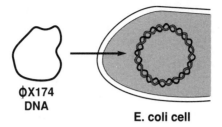

**φX174
DNA**

E. coli cell

FIGURE 6-9
The DNA of phage φX174, a single-stranded circle, is converted promptly
to a double-stranded circle by enzymes present in the infected *E. coli* cell.

Another class of tiny phages which was to have a profound influence on my later work came to me by way of Sankar Mitra. A
native of Calcutta, he joined my laboratory in 1964 for postdoctoral
training after graduate work at the University of Wisconsin. He
brought with him a filamentous virus called M13 (Fig. 6-10), a slight
variant of phage fd discovered a few years earlier by Norton Zinder
at the Rockefeller Institute. These extremely long viruses attach to
tiny tubes (pili) on the surface of *E. coli*. Cells with such pili (called
"male") use them in mating with cells that lack them in a sex act that
transfers some DNA to the "female" recipient. Beyond the fascinating question of how a virus tethered to a pilus injects its DNA into
the cell to start the infectious process was the small size and single-
strandedness of the DNA, features resembling those of phage φX174
DNA. Was the M13 DNA circular too? Indeed it was, and we could
also show that it served as a template for our DNA polymerase as
well as φX174 DNA did.

For many years we tried in vain to find the start of the synthetic
DNA chains. When we learned that DNA polymerase was able to
replicate a single-stranded circle of DNA (a template without ends to

Coat
proteins   DNA

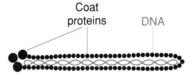

FIGURE 6-10
A filamentous phage (M13) with its single-stranded circular DNA folded
like a rubber band within the protein coat. The three larger protein
molecules at one end moor the phage to a cellular attachment site.

serve as primers), it seemed we had the evidence that the enzyme could start a new chain. Only one barrier separated us from performing the crucial experiment of determining whether the chain we made could generate new viruses upon entering a cell. (Ironically, our later studies showed that neither this enzyme, nor any other DNA polymerase in nature, can initiate a chain; see Chapter 7.) Yet this mistaken confidence that the enzyme could start a chain was what we needed to sustain us through the long and difficult effort of demonstrating the biologic activity of synthetic DNA.

The barrier that stood in the way of our synthetic product's becoming infectious was the requirement that it be circular. Unlike DNA from the virus, our product was a linear chain and would need to have its ends joined to be capable of infecting E. coli. Many of us believed that an enzyme must exist in E. coli that can join or ligate properly abutted ends of a DNA chain to create a circle, as well as to seal breaks in DNA. In 1967 five groups each reported their independent discoveries of such an enzyme, eventually called DNA ligase. Martin Gellert at the National Institutes of Health was first, having used as his assay the capacity of the enzyme to convert the linear duplex DNA of a phage called λ (lambda) into a circular duplex, just as was known to happen immediately upon λ infection of an E. coli cell. Using different assays, and only months behind, were other groups headed by Charles Richardson at Harvard Medical School; Robert Lehman, my laboratory neighbor at Stanford; Jerard Hurwitz at Albert Einstein Medical School; and myself, with Nicholas Cozzarelli as a postdoctoral fellow.

With ligase in hand to circularize a linear DNA and with the Sinsheimer group eager to assay the infectivity of our products, the stage was set for me and Mehran (Mickey) Goulian, my postdoctoral fellow, to determine whether our DNA polymerase could make biologically active DNA from the four nucleotides. Might there be, as some claimed, an infrequent but essential modification of A, T, G, or C, or a novel branching or linkage, beyond the limits of everyone's analyses? Could we rely on the fidelity of DNA polymerase to copy chains of over 5,000 units without a single mistake or omission in matchings of A with T and of G with C?

Goulian did not seem at first suited for the effort. Trained in internal medicine at Massachusetts General Hospital and certified as a hematologist, he had spent two years in African mission hospitals before deciding to do research in biochemistry. At 34 years old, he had relatively little laboratory experience, and I almost made the blunder of not accepting him into my research group. As it turned out, Goulian proved to be a gifted and effective experimentalist. Few matched him in zeal and skill. He was the last to leave the laboratory at night and the first to arrive in the morning. He did manage to see his wife and children during an occasional picnic lunch on the lawn outside the school.

Goulian was a remarkable miniaturist. Once, when I asked him for a reagent and was promptly given a small tube, I could see no solution whatever in it. He protested there was some and proved it by withdrawing into a fine pipet three or four microliters (millionths of a liter), a droplet practically invisible to a normal eye. Goulian went on to practice medicine as a chief of hematology but foremost remained a dedicated basic researcher in DNA replication, first at the University of Chicago and now at the University of California in San Diego.

Our plan for synthesizing infectious viral DNA was carried out in four stages (Fig. 6-11).

Stage I.   We used the single-stranded circular ϕX174 DNA chain (called V for viral) as the template; it was labeled with radioactive hydrogen (^{3}H). With DNA polymerase, we synthesized a matching chain (called C for complementary); it was labeled with radioactive phosphate (^{32}P). We also made our synthetic chains "heavy" by replacing the T building block with an analog in which a bromine atom with a mass of eighty replaces the T methyl group (carbon with three hydrogens) with a mass of only fifteen. (In replication, the enzyme does not distinguish the natural T from this analog because the bromine substitution is on the side of the molecule that is not engaged in its matching with an A in the template.) After the C chain was made circular with ligase, we isolated the duplex product (VC) by centrifugation and examined it in the electron microscope. VC was indistinguishable from the viral duplexes isolated from ϕX174-infected cells. In all its other physical properties, except for identifying labels we had attached, the semisynthetic VC was the same as the natural.

Stage II.   We now wished to isolate the C chain we had synthesized as an intact circle. To do so, we exposed the VC product to DNase, an enzyme which makes random breaks in the DNA chain. The amount

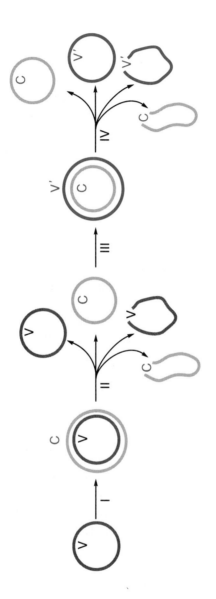

<small>FIGURE 6-11</small>
Stages in producing a synthetic viral DNA in the test tube. The circular, complementary copy, C, of the viral DNA, V (stage I), is isolated (stage II) and used (in stage III) as a template to produce a viral circle that upon isolation (stage IV) can be tested for infectivity.

of enzyme and length of exposure was adjusted such that on average there was only one break per duplex, making one chain linear and leaving the other intact. Thus, roughly half of our synthetic C chains should have survived as circles. How were we to harvest them?

The techniques had been devised ten years earlier by the late Jerome Vinograd at the California Institute of Technology. By including heavy atoms in the DNA chain, he learned that he could separate the heavier chains from the lighter ones in a centrifuge. Intact circles could also be separated from broken ones in the centrifuge because they too sediment more rapidly. Thus, after melting the chains of the duplex (VC) circle apart by heating, we applied the Vinograd methods to separate the different forms of DNA chains. Having unequivocal means to trace them, we isolated the synthetic (^{32}P-labeled) C circles and could now assess their infectivity. To our great relief and delight, they were infectious. They were as potent as C circles prepared the same way from VC molecules produced naturally in infected cells.

Stage III. Now that we had isolated the synthetic C circles, we could use them as templates for replication. The operations were as in Stage I, except that the natural T rather than its brominated analog was used. We now had produced an entirely synthetic VC duplex (Fig. 6-12).

Stage IV. The final test! We applied the maneuvers of Stage II to this fully synthetic duplex. From the mixture of synthetic chains (Vs and Cs, broken and circular) we could isolate V circles with the physical properties of the chains obtained from φX174 viruses. They were just as infectious!

After twelve years of trying, we had finally done it—we had gotten DNA polymerase to assemble a DNA chain with the identical form, composition, sequence, and genetic activity of DNA from a natural virus. All the enzyme needed was the four common building blocks: A, G, T, and C. At that moment, it seemed there were no impediments to the synthesis of DNA, genes, and chromosomes. The way was open to create novel DNA and genes by manipulating the building blocks and their templates.

In a very small way, we were observers of something akin to what those at Alamogordo on a July day in 1945 witnessed in the explo-

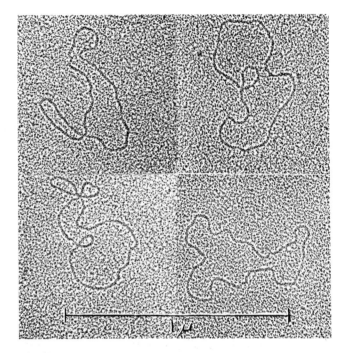

FIGURE 6-12
Electron microscopic picture of the totally synthetic dou-
ble-stranded DNA (stage III in Figure 6-11). (Courtesy of
Dr. Jack Griffith.)

sive force of the atomic nucleus. Harnessing the enzymatic powers of
the cellular nucleus had neither the dramatic staging of light and
sound of the atomic bomb nor the stunningly apparent global consequences. Yet, this demonstration of our power with enzymes that
build and link DNA chains would soon help others forge a different
revolution, the engineering of genes and the modification of species.

Creation of Life in the Test Tube

A hundred newspaper and television reporters and photographers
came to a press conference called by Spyros Andreopoulos, Director
of the Stanford News Bureau on December 14, 1967, because of
many inquiries about a paper we had just published in that month's
issue of the *Proceedings of the National Academy of Sciences.* The

title was "Enzymatic Synthesis of DNA, XXIV. Synthesis of Infectious Phage φX174 DNA." To the editors who sent the newsmen, it seemed that a virus had been synthesized in the test tube, and they thought this a newsworthy event.

After the conference, I overheard a reporter on the telephone to his office saying: "It's not what we expected. They haven't made a virus. It's only a molecule, a short chain of DNA. They've been making DNA in the test tube for twelve years." Hairy little monsters had not been created in the test tube. On the other hand, a fully infectious viral chromosome had been assembled from the four synthetic nucleotide building blocks. As a biochemical landmark, it was newsworthy. This twenty-fourth paper in a series on the test-tube synthesis of DNA proved that a chromosome is a linear chain of the known nucleotides with no extraneous components, and that the enzyme DNA polymerase, discovered twelve years earlier, can copy a preexisting viral DNA chain of 5,000 or more nucleotides, error-free. For these reasons, DNA polymerase would become a major tool in the still unborn genetic engineering revolution.

Perhaps the editors and reporters were excited about the test-tube synthesis of a viral DNA because they did not understand the difference between DNA and a virus. Perhaps also viruses, as agents of diseases for which there were few effective medical interventions, seemed to them far more formidable creatures than the thousand-times-more-complex bacteria that succumb to antibiotics. No wonder the reporters besieged me with the question: "Is the DNA you've made a 'living molecule' "?

At the news conference, I tried hard to explain why the answer to that question is so elusive. Viral DNA has no life of its own, nor does the virus that bears it, in the sense of being able to grow and reproduce outside the cells of an organism. Yet the DNA has a potent life force. Once inside a bacterial, plant, or animal cell, viral DNA can divert the host machinery to doing little else but making thousands of viruses identical to itself. Across the wide gamut of viruses, some are large and complex, so much so that they come close to resembling the simplest of cells. Are cells alive? No one would question that. Yet there are cells so stripped of walls and metabolic equipment that their survival is tenuous without the support and nourishment of neighboring cells. I tried to make it clear that, in essence, there is no point at which the breath of life is infused in ascending from carbon atoms to nucleotides to DNA to viruses to cells to man.

Andreopoulos, whose notes on events of that busy day add elements which I had either forgotten or never knew, recorded that "from the start of our plans to make a public announcement with the simultaneous publication of the work in *PNAS*, you had asked that the phrase 'Synthesis of Life in the Test Tube' be avoided. I spent as much time as you did with reporters cautioning against characterization of the work in those terms. Then in a telephone conversation, science reporter Harold Schmeck of *The New York Times* suggested the phrase, 'Synthesis of the Inner Core of a Virus.' It wasn't particularly great, but less objectionable."

Despite our disclaimers, there was still enough excitement left that day about creating "life" in the test tube to make front page stories in virtually all newspapers worldwide. Attention was heightened by remarks President Lyndon B. Johnson had made that day at the Smithsonian Institution in Washington, at a ceremony observing the bicentennial of the *Encyclopaedia Britannica*. Andreopoulos wrote:

> On the morning of our scheduled press conference, I received a call from a speech writer in the Office of Science and Technology. The writer asked, "Could we draw a few paragraphs for inclusion in the President's Smithsonian speech on December 14?"
>
> I phoned you immediately about this. Your initial reaction was, "Who do we know in government who can stop this?" I said the suggestion to the President had come from NIH Director Jim Shannon in response to the Senate Appropriations Committee which earlier had chastised the NIH for failing to publicize the federal role in health research. You decided to comply.
>
> That same afternoon your press conference took place at Stanford. You explained the various steps involved in the research and assessed its potential. You also made a big point in saying that there was some disagreement among scientists as to whether viruses were "life."
>
> That evening I went home and turned on the TV. The lead story on the 6 o'clock news was the DNA synthesis. ABC anchor Roger Grimsby began by saying, "In reporting developments in science, there are times when we journalists can be accused of oversimplification. We will switch to Washington to let President Johnson tell you what happened at Stanford today."
>
> When he appeared on the screen, my anticipation was heightened. LBJ began to read the first line of the prepared statement, "They have for the first time succeeded in manufacturing a synthetic molecule . . ." Here LBJ paused, put the page aside and said, "What are you going to read about tomorrow morning? It is going to be one of the most important stories that you ever read, your Daddy ever read, or your Grand-

pappy ever read . . . Some geniuses at Stanford University have created life in the test tube!"

The next day every news account led with that headline, plus LBJ's prepared statement.

That statement read as follows:

At this very moment, biochemists at Stanford University are announcing a spectacular breakthrough in human knowledge. They have for the first time succeeded in manufacturing a synthetic molecule in a living organism. In their work, they have come closest yet to creating life in the laboratory by manufacturing the living genetic material of a virus. When this man-made viral material infected a bacteria, it began to reproduce itself.

These men have unlocked a fundamental secret of life. It is an awesome accomplishment. It opens a wide door to new discoveries in fighting disease and building healthier lives for mankind. It could be the first step toward the future control of certain types of cancer.

The work of these scientists headed by Dr. Arthur Kornberg is living proof of the creative partnership which has developed over the years between science, the universities and the federal government. We are proud that our explorations have been made possible by grants from our National Institutes of Health and National Science Foundation.

As man continues to make these astonishing new discoveries, I devoutly hope that he will also grow in the wisdom needed to apply them for the benefit of all mankind.

Ironically, Johnson had already begun the deceleration of government support of basic research that was to set a pattern for subsequent administrations.

Editorial reactions were uncomfortably effervescent.

New York Times: "The perspectives opened up here are as breathtaking as those exposed several decades ago when biologists first began using marked atoms made artificially radioactive so that the fate of different compounds in living organisms could be studied directly."

Chicago Sun-Times: "It was a victory for pure research—exploration for the sake of exploration even though there may never be a practical application of knowledge that is uncovered."

Los Angeles Times: "The history-making synthesis not only may give us new insight into understanding virus infections, but also may open up new avenues of research in finding out what takes place when normal cells are changed into malignant, cancerous cells."

Time: "Since the dawn of science, one of mankind's most impossible dreams has been the creation of life in a test tube. Last week, scientists moved a step closer to making the dream possible."

The story I liked best was written for the *Manchester Guardian Weekly* by Alistair Cooke, to this day my favorite TV commentator and essayist. His dispatch read:

San Francisco, December 17

Kornberg, a name that biologists and biochemists conjure with, has suddenly found itself in the newspapers of the world in the company of such names as Koch, Ehrlich, Pasteur, and Einstein.

Its owner is pretty uncomfortable about this, and when he entered the Tresidder Memorial Union, at Stanford University near here, and heard the clatter of television crews and saw a nucleus of microphones and an army of respectable newsmen, his fears of premature immortality were fulfilled.

He was sanctified overnight last week for having reported to the National Academy of Sciences that he and two colleagues had successfully synthesised a biologically active virus DNA, the chemical of all heredity. "Life," the press reported, "has been created in a test tube." It is near enough to the truth to astound the layman, far enough away to annoy the expert.

Dr. Kornberg appeared before the laity with the responsible purpose of giving honour where honour is due, setting the record straight, and trying, with absolutely unruffled calm, to make vital distinctions which for most of us seemed like a semantic experiment in redefining a hair in order to split it.

Had he, to begin with, really "created life in a test tube"? He took off his dark-rimmed glasses, breathed a living sigh and pondered. "Well," he said, "it is really impossible to define life and living to the satisfaction of both laymen and scientists. There is a lively controversy about whether a virus is alive or not. Bacteria are alive. Perhaps most scientists believe that viruses are too."

What he and Dr. Mehran Goulian, of the University of Chicago Medical School, had done was to create the living core material of a virus that reproduces and makes new viruses which live like other viruses. "With the reservations I've mentioned, it becomes reasonable to think of the viral DNA as a simple or primitive form of life."

The new synthesis followed his procedure of 11 years ago. Up to the last step, the long experiment had repeated precisely the feat that earned him the Nobel Prize in 1959: the synthesising of a virus which resembled living virus DNA. But only in the last year had it been possible, by the discovery of an enzyme, to take the vital step of making the synthetic DNA able to reproduce itself.

With excruciating patience and exquisite clarity (I guess) he described the process of packing nucleotides (the building blocks of life, silly) into a test tube with a familiar enzyme, DNA polymerase, and "radioactively tagged template of natural viral DNA." Polymerase, it will be no surprise to learn, used the template as its programme to direct and speed the non-living nucleotides into a long chain of viral DNA, "an exact copy of the template material." The tagged natural DNA was then removed. The new "joining enzyme, DNA ligase" was added. By this means, the two ends of the chain of molecules were joined into a circle.

Kornberg and Goulian had done it! And they knew why the old synthetic DNA had failed to reproduce itself: the ends of the molecular chain were not joined.

The two men sent the new viral DNA to Dr. Robert Sinsheimer at California Institute of Technology, because he is the man who discovered and identified PhiX174, which is the name of the natural virus used as a template. Dr. Sinsheimer reported that the new DNA was indeed biologically active. Other scientists, hearing this report, are asking whether Kornberg and Goulian really did remove all the natural template DNA from the test tube.

Having said what had been done, Dr. Kornberg was charmingly anxious to say who should take the credit, and then to reassure some panicky people about a problem in ethics.

First, he would say "we know scandalously little about DNA chemistry, we are only at the beginning of a long programme of basic research." But even so, this beginning did promise "the creation of genetic material with hitherto unknown characteristics" and the creation of "mock cancer viruses" that might give immunity against the pathological virus.

At the end, the moral problem was posed. "Dr. Kornberg, do you see the time when your work will come into conflict with traditional morality?" Again he took off his glasses and looked down and meditated.

Very gently, he replied: "We can never predict the benefits that will flow from advancements in our fundamental knowledge. There is no knowledge that cannot be misused, but I hope that our improved knowledge of genetic chemistry will make us better able to cope with hereditary disease. I see no possibility of conflict in a decent society which uses scientific knowledge for human improvement."

Not all the reactions were favorable. Some cartoons pictured laboratory scenes in which startled scientists see homunculi emerging from their test tubes (Fig. 6-13). Max Perutz, the eminent British molecular biologist, in a letter to the London *Times*, complained

"I don't understand it either!"

FIGURE 6-13
A newspaper cartoon of December 15, 1967.

about the fuss over a discovery that was anticipated, and one whose practical significance he doubted. I agreed with him about the excessive publicity and, like him, failed to anticipate that this discovery and related ones to follow would lead rapidly to recombinant DNA and the extraordinary applications of genetic engineering.

A newspaper story on January 4, 1968, seemed to grasp the significance that Max and I missed. It was headlined: "Creation of Life Rated Best of Science Stories in 1967." In smaller type: "Human Heart Transplant Second."

Chapter 7

Astonishing Machines
of Replication

"I'll bet you a bottle of champagne," said Buzz Baldwin, a colleague in my department, "that the nuclease (DNA-degrading) activity in DNA polymerase is part of the enzyme." The preparation of the replicating enzyme we had purified extensively could still degrade DNA chains. In the absence of the nucleotide building blocks needed for synthesis, nucleotide units were cleaved slowly and serially from DNA. I took Baldwin's bet because it made no sense to me at the time that DNA polymerase would degrade the very end of the chain it would normally be extending. The bet also gave me the incentive to purify the polymerase further and remove the nuclease and other putative trace contaminants. Two years later, I accepted the nuclease activity as an integral part of DNA polymerase, understood why it was there, and paid off the bet.

To begin with, we were unable by several procedures to reduce the activity of the nuclease relative to the polymerase activity. The ratio of the two activities remained constant throughout the last stages in which the enzyme was purified 100-fold relative to other proteins. Then a simple fact about the nuclease gave us our best clue. Douglas Brutlag, working on this problem for his graduate thesis, observed that the degrading activity was far more potent on a single strand of DNA than on the usual double-stranded form. This preference became extreme when the temperature of the reaction was lowered. At

20°C, degradation of single strands was reduced only moderately from what it was at 37°C, but that of duplex DNA was nearly abolished. Presumably, the ends of duplex DNA are slightly frayed (melted) at 37°C, but locked together (frozen) and therefore never appear single-stranded at the lower temperature. Why should a loose primer end be a substrate for degradation by a synthesizing enzyme?

Proofreading and Editing by a Replicating Enzyme

Rarely does a series of experiments give so clear a picture. Brutlag prepared a variety of duplex DNAs in which the primer end of a chain was not matched to the other strand and thus was frayed at all

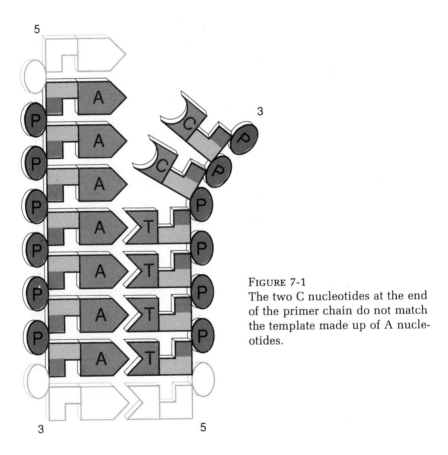

FIGURE 7-1
The two C nucleotides at the end of the primer chain do not match the template made up of A nucleotides.

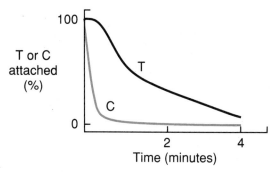

FIGURE 7-2
The mismatched C nucleotides are removed promptly from
the primer chain, followed by a more gradual removal of
the T nucleotides.

temperatures. One such duplex was a short chain of Ts aligned opposite a long chain of As; the primer end of the T chain had a few Cs
attached which could not pair with the As (Fig. 7-1). The Ts were
labeled with radioactive phosphate; the Cs were distinguished by
being tagged with radioactive carbon. Upon exposure to DNA polymerase, all the mismatched Cs were removed within a minute, after
which the Ts were removed far more slowly (Fig. 7-2).

The very same experiment was repeated, but this time T building
blocks were included in the reaction to support extension of the
chain. What a beautiful result! All the Cs were removed quickly just
as before, but the chain of Ts remained intact and was immediately
extended by synthesis along the template A chain (Fig. 7-3).

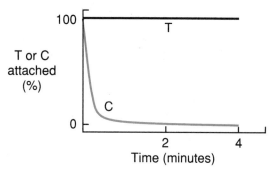

FIGURE 7-3
The mismatched C nucleotides are removed promptly, but
the T nucleotides are protected by synthetic addition of
new Ts, which match the template.

From many such experiments, it became apparent that the enzyme removed all the mismatched units. Fresh units were added to the primer-chain end only when the end of the primer was correctly matched to the template chain (A to T; G to C). We could properly infer that if the synthesizing enzyme were to make a rare mistake during elongation of a chain, such as inserting a C opposite an A (estimated to happen once in ten thousand times), it would remove the mismatched C before proceeding with extension of the chain. This astonishing *proofreading* ability of the enzyme, coupled with its fine discrimination in the choice of correct building blocks during synthesis, reduces errors in the overall process of replication to one in ten million.

A vivid demonstration of this proofreading mechanism came from Maurice Bessman, who was exploring the properties of another DNA polymerase. In his postdoctoral fellowship with me in St. Louis, Bessman had made many key contributions to the earliest work on DNA polymerase. He then took a position in the Biology Department at Johns Hopkins University and began to study the novel DNA polymerase produced in *E. coli* during infection with the phage T4. (This phage, twenty times more complex than the tiny φX174, has a set of genes that directs the host cell to produce a unique replication apparatus to suit the special needs of the phage for rapid DNA synthesis.) Among mutants of this phage, some have an alteration in the gene for DNA polymerase resulting in an enzyme with different properties, depending on just where and how the gene has been altered. Some of these mutant phages make more mistakes in replicating their DNA and hence were called mutator mutants; others that make fewer mistakes than normal were called antimutators. Bessman found that in most mutators the proofreading nuclease activity of the polymerases was weaker than normal, whereas in the polymerases of the antimutators, the nuclease activity was exaggerated. In other words, fidelity of phage DNA replication was nicely correlated with the proofreading capacity of the DNA polymerase.

Having finally made sense of why an activity that degrades DNA is part of the very enzyme that makes it, we were nevertheless unprepared for the next shock. Bob Lehman made the paradoxical observation that the nuclease activity of DNA polymerase on double-stranded DNA was enhanced ten-fold when all four building blocks required for synthesis (A, T, G, and C) were present. How could

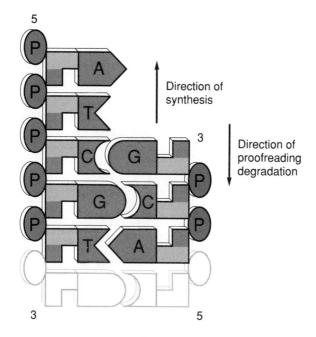

FIGURE 7-4
A DNA chain might be susceptible to stepwise degradation
from either end, the primer (3) end or the other (5) end.

synthesis be enhancing degradation? After all, we had observed earlier that synthesis extends the primer end of a chain and thereby
protects it from nuclease action.

Keys to the solution came from two laboratories. At the Rockefeller Institute, Edward Reich and his colleagues observed that an
analog resembling A was added to a DNA chain by DNA polymerase
but not removed by the nuclease activity of the enzyme. Yet the
radioactively labeled chain to which the analog was added was
steadily degraded. In my laboratory, Murray Deutscher, a postdoctoral fellow, found that a chain with its primer end modified by
attachment of a phosphate group was inert for synthesis by DNA
polymerase, but nonetheless was degraded by the enzyme.

To interpret these observations, we must first recall the backbone
structure of a DNA chain. In the brief segment shown in Figure 7-4,
the OH group attached to carbon 3 is the primer end (growing point)
of the chain, the end at which proofreading also occurs. The phos-

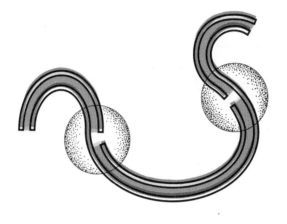

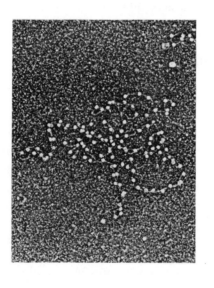

FIGURE 7-5
DNA polymerase binds to a nicked region in a DNA chain, as shown in the electron microscopic image and the diagrammatic sketch.

phate groups (P) connect carbon 5 of one sugar to carbon 3 of the next in the chain. The chain has two distinctive ends, called 3 and 5.

We reasoned, as did the Rockefeller group, that there must be an additional nuclease activity in the DNA polymerase preparation that degrades serially from the 5 end of the chain and is not affected by blockage of the 3-end by an analog or phosphate. By several devices, including labeling of the 5-end with radioactive phosphate, we could show plainly that this was true and concluded that the enzyme must possess a separate domain designed for this particular function.

We could now explain the Lehman paradox: how the four building blocks, and the synthesis they make possible, enhance nuclease action from the nongrowing 5-end of the chain. The answer lay in the DNA we were providing the enzyme. DNA is vulnerable to occasional breaks in the backbone of one of the chains. We could locate them in the electron microscope (Fig. 7-5) because DNA polymerase finds such nicks, and we could see where it had latched onto them. Synthesis cannot take place at a nick because there is no template free for the building blocks to pair with (Fig. 7-6). Only upon removal of DNA from the 5 end of the nick does a stretch of template become exposed for pairing with a substrate nucleotide and synthesis. In its synthetic progress, the polymerase slides along the template (from left to right) and is thereby brought up to a 5 end of the chain, which it then degrades. In this manner, synthesis stimulates removal of DNA from the 5 end. Is there any point to this apparently futile exercise by DNA polymerase in removing DNA from the 5 end, thus creating a gap, while filling it by extending the 3 end? Emphatically so. Some years later, we would recognize that this maneuver by polymerase is an essential step in replication. What was obvious immediately was how this nuclease action by polymerase might give it a crucial role in one of the most basic biologic operations, the *repair* of lesions in DNA.

In all cells in nature, from the most primitive to the highly specialized, several devices are at instant readiness to detect and correct the variety of lesions to which its precious DNA is prey. Damage by cosmic rays and ultraviolet rays, by chemicals in food and air, by oxygen activated in metabolism and the errors introduced in replication, all must be repaired to preserve the integrity of the master molecule upon which the survival of the organism and species depends. The lesions caused by ultraviolet (UV) light are among the

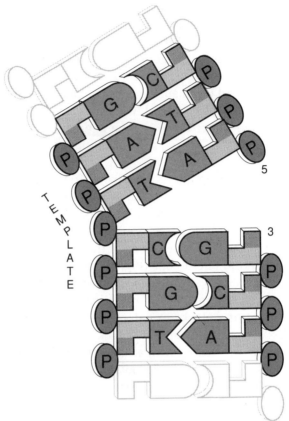

FIGURE 7-6
At a nick in the DNA chain, with no template exposed, extension of the primer (3) end is possible only if the 5 end to
which it abuts is degraded away.

most common and pervasive. A UV ray impinging on two Ts in a
row in a DNA chain merges them into a linked structure (Fig. 7-7) so
that each can no longer pair with an A in an opposite chain. If
unattended, this aberration will either obstruct replication or cause a
mutation at that spot. The UV lesion must be removed, and we
thought DNA polymerase might have a key place on the team of
enzymes that does it.

The DNA of a skin cell in sunlight suffers 50,000 UV lesions during every hour of exposure. The DNA of a bacterial cell is just as
susceptible. To cope with this withering barrage in a one-hour sun-
ning, estimated to increase cancer risk 10,000-fold, enzymes have

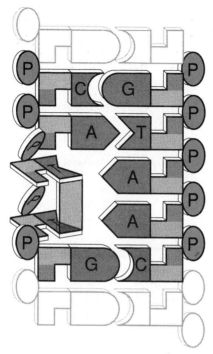

FIGURE 7-7
Exposure to ultraviolet light causes neighboring T nucleotides
to become linked, thus distorting the chain and preventing the
normal pairing with A nucleotides.

evolved that patrol the vast length of the chromosome, feeling the
DNA backbone for the molecular distortion of a pair of merged Ts.
At the site of such a lesion, one of the repair enzymes breaks the
backbone, marking it for excision (Fig. 7-8). DNA polymerase, which
binds avidly to any nick in a chain, removes the damaged section
(and commonly more of the chain for good measure) and then fills
the gap using the opposite chain as a template. Finally, another
member of the enzyme team, ligase, the joining enzyme, seals the
break in the backbone, leaving the DNA without a trace of the initial
damage.

Studies from several laboratories, including our own, showed con-
vincingly that this cut, patch, and seal operation goes on in all kinds
of cells; in E. coli, it was the DNA polymerase we had become familiar with which could excise the UV lesion and replace it with
healthy DNA.

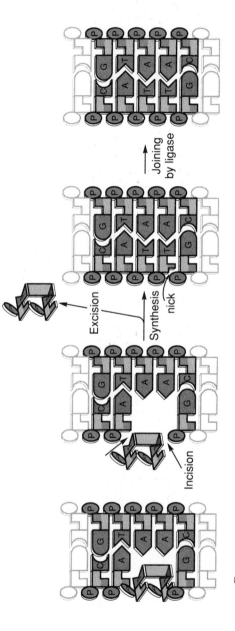

FIGURE 7-8
DNA damage caused by ultraviolet light is repaired by removing the patch containing the linked T nucleotides and replacing it with proper DNA by synthesis with DNA polymerase.

DNA Polymerase under Indictment

"DNA polymerase, you are charged with masquerading as a replication enzyme. Your ability in repairing DNA damage has been misrepresented by your agents as relevant to replication. You are a red herring." Such, in paraphrase, were the accusations against DNA polymerase and me by *Nature New Biology* in a series of unsigned, defamatory editorials in 1970 and 1971.

The replicative role of DNA polymerase was questioned for several reasons. First was the versatility of the enzyme in excising a lesion from DNA and patching the damaged molecule. Then there was the Cairns mutant! In 1969 John Cairns had found a most remarkable mutant of *E. coli* which appeared to lack the enzyme and yet it grew and multiplied at a normal rate. He had discovered the mutant by a "brute-force" effort in which extracts made from thousands of colonies (derived from a culture exposed to a near-lethal level of a mutagenic chemical) were assayed for DNA polymerase activity. An extract of the 3,478th colony examined had only 0.5 to 1 percent of the normal level of enzyme. This strain, impaired only by a heightened sensitivity to ultraviolet light, became the star witness for the prosecution.

In addition to the estimable qualifications of DNA polymerase for repair of DNA and its apparent dispensability for cell multiplication, other evidence was accumulating against a principal role for DNA polymerase in replication. Genes were being discovered (designated *dnaA*, *dnaB*, *dnaC*, and so on) which strongly implicated many other proteins as essential for a replication process far more complex than had been imagined.

I might digress to describe the clever way of obtaining mutants which, though defective in a vital function such as replication, remain viable. First introduced by Norman Horowitz in studies of *Neurospora* at the California Institute of Technology, the method has been widely applied to bacteria and animal cells. A cell or organism unable to make an essential small molecule (such as adenine or vitamin C) can circumvent this deficiency by importing the molecule from the nutrient environment. But it is impossible to feed a cell a large molecule, such as an intact chromosome or the enzyme needed to make it. How then can one obtain a viable mutant with a defective replication enzyme? The technique depends on selecting those mutations in an essential gene which enfeeble but do not completely destroy the function of the protein encoded by that gene. With certain mutations in the *dnaA* gene, for example, the resulting proteins can sustain replication at the mild temperature of 30°C (77°F) but fail at the harsher temperature of 42°C (108°F). Many such thermosensitive mutants can be prepared and are referred to as "conditionally lethal," because their survival depends on a condition in the environment (say a milder temperature) that accommodates their deficiency.

Despite the mounting evidence against DNA polymerase as a replicative enzyme, I was not ready to enter a plea of guilty. Possibly the low activity of polymerase in the Cairns mutant might be due to poor extractability of the enzyme rather than its absence in the intact cell. Perhaps the enzyme, though defective, could still function inside the cell but not in the artificial environment of an extract. Conceivably, the normal abundance of the enzyme might be so generous that a residual 1 percent suffices to sustain replication.

The rising skepticism about the importance of DNA polymerase was fanned by the *Nature* vendetta. Not only was the enzyme attacked but the basic mechanism, the building blocks, and the assays we used to measure DNA synthesis were judged to have misled a generation of biochemists and were now impeding the discovery of the true DNA replicating enzymes. At this juncture, my son Tom entered the fray. It was May 1970, when I was at the Molecular Biology Laboratory in Cambridge, England, in the second half of a sabbatical year devoted to learning more about cell membranes.

The reason I was there is that I had felt uneasy about basic questions regarding membrane chemistry that had arisen during the previous six years' work on the production and germination of bacterial spores. More immediately, I wondered whether our failure to discover how DNA chains are started in replication might be due to inattention to membranes. For these reasons, I had set out to learn about physical and chemical methods practiced in some of the leading membrane laboratories: those of Harden McConnell at Stanford, George Palade at Rockefeller University in New York, Alec Bangham in Cambridge, England, and Laurens van Deenen in Utrecht, Netherlands. At the end of a pleasurable year (including motor trips around England, Scotland, France, Spain, and Holland), I decided that in my future work I would continue to focus on replication and look for proteins that might operate from a membrane base, rather than switch to studies of the phospholipid components of membranes. I also wanted to get back to the basics of replication enzymology.

But in May, while I was still in Cambridge, Tom had called from New York for two reasons. Swelling in his left index finger had worsened, making it impossible for him to continue his cello training at the Juilliard School. Second, he was distressed over disparaging comments about DNA polymerase in his biology course at Columbia College, where he was also a full-time student. Since he was unable to play the cello, Tom wondered whether he might search for the missing polymerase in the Cairns mutant.

Unlike both his brothers, Tom had no childhood or later experience working in the laboratory. "Don't bother," I told him, "Charles Richardson's been trying, but even with his expertise he hasn't found any novel polymerase activity in the Cairns mutant." Not to be discouraged, Tom obtained bench space in Malcolm Gefter's laboratory in the Biology Department at Columbia and within three weeks had found a DNA polymerase in *E. coli* cells distinct from the one I had discovered. In September he presented his sensational findings at the International Congress of Biochemistry at its triennial meeting in Switzerland, barely three months after he first entered a research laboratory.

During the next year, as a graduate student, Tom purified the new polymerase activity and named it DNA polymerase II (pol II). He could clearly distinguish it from the "classic" one now known as DNA polymerase I (pol I). In the course of a chromatographic separation, he noted still another band of polymerase activity emerging quite separately from pol II and in a position which would have been occupied and obscured by pol I in preparations from normal cells. Despite strong opinions from experienced enzymologists that this new band was likely a technical artifact, he persisted and proved that it was a distinctive entity, which he named DNA polymerase III (pol III). Subsequently, he and Gefter located the gene for pol III and showed that conditionally lethal mutations in this gene blocked DNA replication. Pol III, in a far more elaborate form that still defies a complete description after fifteen years of intensive study, was to gain recognition as the keystone of DNA replication in *E. coli*.

All three polymerases, although differing significantly in structure, proved to be virtually identical in their mechanisms of DNA synthesis, proofreading, and use of the same building blocks. The maligned polymerase (pol I) became the prototype for all DNA polymerases in plants, animals, and viruses, as well as in E. coli. The gloomy prophecies of Nature New Biology soon disappeared, as did the magazine itself.

Following discovery of the Cairns mutant, other studies had soon showed that a mutation in any one of a dozen genes would severely impair replication. Identifying a gene and locating it on the chromosome map is a far cry from knowing the function of the protein it encodes. The replication process, which we had once thought was performed solely by two enzymes, DNA polymerase and ligase, was obviously far more complicated. How were we to find the many proteins that would then have to be fitted together to constitute the intricate machine of replication?

DNA Polymerase Cannot Start Chains

Despite the excitement over the synthesis of a chain of infectious viral DNA back in 1967, I had felt a certain uneasiness. One of the inferences drawn from the replication of a single-stranded, circular template was that DNA polymerase I could start a new chain. Yet we were never able to find direct proof of this. The first building block to start a chain should retain its activated triphosphate form (Fig. 7-9), whereas all subsequent additions to the chain lose their terminal activating group as P-P. Even with intensive radioactive labeling, we failed to detect any trace of an initiating unit.

Moreover, we had observed that replication of the circular template was far more efficient if a small amount of boiled E. coli extract was present. Why? Perhaps this juice contained fragments of DNA, which by direct extension would prime the growth of a new chain. Yet it seemed unlikely that random fragments of DNA would match the viral DNA template accurately enough to serve as primers. In 1969, this unlikely possibility became a reality after we discovered the proofreading and editing capacities of DNA polymerase.

The stimulatory effect of this boiled E. coli extract was demonstrably due to fragments of the E. coli chromosome. We could remove the effect of the extract by a complete digestion with a DNA-degrading enzyme (DNase); also, the extract could be replaced with partially digested (fragmented) E. coli DNA. The sequence of events

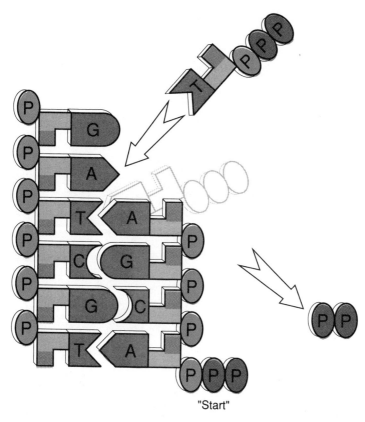

"Start"

FIGURE 7-9
The start of a chain by DNA polymerase should be marked by retention of
the activated (triphosphate) form of the initiating nucleotide.

was now clear (Fig. 7-10). A small part of a random fragment
matches by chance a corresponding sequence in the viral template.
Proofreading by polymerase removes all unmatched units from the
3-end of the fragment. Synthesis proceeds to the extent of available
template. Editing from the 5-end removes not only the unmatched
units but also a generous portion of the matched region; polymerase
fills in the gap. Finally, ligase closes the circle. With radioactively
labeled *E. coli* DNA fragments, we could detect the retention of no
more than one "primer" nucleotide unit for every five circles (27,000
nucleotides) replicated. Our 1967 claim that the synthetic viral circle was made wholly from the four building blocks we furnished the

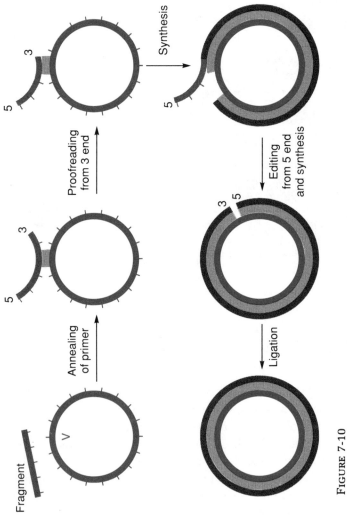

FIGURE 7-10

Replication of viral DNA by DNA polymerase requires an extraneous fragment, annealed as a primer to start a DNA chain; proofreading and editing by the polymerase erases vestiges of the primer.

enzyme was still correct. In nearly every circle, the primer fragment had been eliminated without a trace.

Once again, we were left with the question: Exactly how is a new DNA chain started when a viral template enters a cell? Surely, selection of a primer from a pool of random fragments in the cell is too clumsy and haphazard to be Nature's way. A similar question about starting chains had to be answered for the far more universal circumstance involved in the replication of virtually all chromosomes. Reiji Okazaki had shown four years earlier, in 1967, that chains are started not once, at the beginning of the chromosome, but repeatedly in staccato fashion during the progress of replication. This discovery had made him famous.

When Reiji died of leukemia in 1975, at the age of 45, newspaper accounts listed him as a casualty of the atomic bomb, because thirty years earlier he had searched for his parents in the rubble of Hiroshima. Since the incidence of leukemia at the time of his death was as frequent in the general population as among those exposed to the radiation, the source of his illness must remain uncertain. What was painfully and tragically clear, however, was that the world had lost a gifted experimentalist in his prime. When Reiji and his wife Tuneko left my laboratory in 1963 after a productive postdoctoral research stay, they returned to Nagoya University, where they made their epochal discovery.

Reiji's research style is not readily gleaned from his publications, but it is vivid in my memory. One example I call the Okazaki maneuver. In purifying an enyme, he used a heating step: 10 milliliters of the enzyme solution was held in a test tube at 70°C for five minutes; then the coagulated impurities were removed by centrifugation. When he came to the point of scaling up the procedure to several liters, he simply repeated the original heating procedure several hundred times. I was embarrassed to have to report such an unsophisticated procedure in our publication. But then I realized that he was able to complete this step in a few hours and saw no point in wasting precious time and material in learning how to do the heating in a large beaker or flask. Later, when another of my students purified an enzyme with a heating step and had to scale up his procedure 2,000-fold from 3 milliliters to 6 liters, I advised the Okazaki maneuver: he added 3 milliliters to each of 200 test tubes to carry out the heating and centrifugation step, repeating the procedure nine times. Another student who tried heating a large volume of this enzyme in a single step lost it in a thick coagulum.

Consider a major paradox of DNA replication in 1967: With each of the chains of the parental duplex serving as a template, two new chains are assembled, thereby generating two daughter duplexes identical to each other and to their parent. At a gross level, as seen in the microscope or by genetic analysis, progress of the replication fork appears as a synchronous growth of both daughter duplexes (Fig. 7-11). However, at the molecular level, concurrent synthesis of daughter chains on both parental template chains was incongruous.

Recall that the two chains of a duplex are oriented in opposite directions and that synthesis by DNA polymerase proceeds only in the 5→3 direction, a direction opposite that of its template (Fig. 7-12). How could the evidence for synchronous growth of *both*

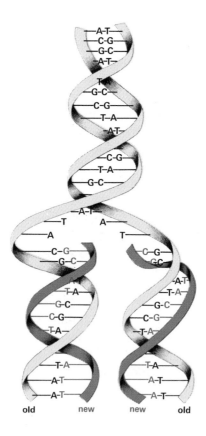

old   new   new   old

FIGURE 7-11
Replication of double-stranded DNA advances at a fork, as proposed by Watson and Crick.

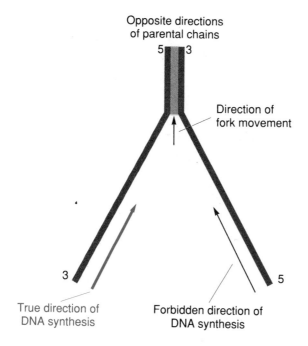

Opposite directions
of parental chains

5   3

Direction of
fork movement

3   5

True direction of
DNA synthesis

Forbidden direction of
DNA synthesis

FIGURE 7-12
Chain growth of the two daughter strands cannot be in the same direction
because the strands of the parental duplex are oriented in opposite directions and DNA polymerase can copy only in the direction opposite to its
template.

daughter chains be reconciled with this limitation of DNA polymerase? To resolve the dilemma, we and others proposed a scheme
in which synthesis along the two parental chains progresses in different ways. Along one parental chain, synthesis by DNA polymerase proceeds continuously in the standard way to generate a
"leading chain" (Fig. 7-13). Along the other parental template, new
chains are started repeatedly. The *discontinuous* synthesis of the
"lagging chain" might appear to be continuous because each newly
started chain is eventually united by ligase with the one previously
laid down. In this way, the growth of the lagging strand, although
opposite in direction to that of the main fork movement at the
molecular level, seems at a gross level to be in the same direction.

Okazaki and his colleagues, in remarkable studies of replication of
bacteriophage DNA and the *E. coli* chromosome, showed convinc-
ingly that a significant fraction of newly synthesized DNA is in the

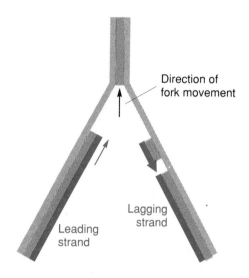

FIGURE 7-13
Advancement of a replication fork by continuous synthesis of a leading strand and the discontinuous synthesis of the other (lagging) strand; the short (nascent) fragment is later joined to the main body of the strand.

form of short pieces, about one thousand nucleotides long. Formed at a rate of about one per second at the fork of the bacterial chromosome, the "Okazaki fragments," as they came to be known, vanish when they are linked to the main chain, which when completed is more than a thousand times as long. The semidiscontinuous mechanism of replication, made credible by the Okazaki experiments, applies to double-stranded animal viruses and chromosomes as well as to the bacterial chromosome. Implicit is the repeated initiation of new strands of DNA for the growth of the lagging strand at the replication fork. The problem of how chains are started loomed like a multiheaded Medusa. And this was not the only problem confronting us.

How DNA Chains Are Started

One day in 1971 Noboru Sueoka, an esteemed friend and colleague from the University of Colorado, began the seminar he had been invited to give at Stanford by saying: "Very little is known about replication." I was not pleased. I would not have minded had he instead enumerated the many questions about replication that remained despite the considerable knowledge we had acquired about its basic mechanisms. How is a DNA chain started? What are the replicative functions of the three distinctive DNA polymerases in *E. coli*? What are the specific functions of the numerous *E. coli* genes identified as essential for replication? What is the switch that controls initiation of a cycle of chromosome replication? Sueoka's comment was not the only annoyance I had endured lately. In a popular cartoon (Fig. 7-14), a fig leaf was placed discreetly over the replication fork to symbolize our innocence about the elements that propel replication.

The psychological goad of this cartoon, added to my frustration with our lack of progress, may have spurred me finally to recognize a basic flaw in our work. It dawned on me that we would never answer the extant questions about replication with the DNA we were using as template and primer; DNA extracted from bacterial and animal cells was not an adequate substrate for the enzymes of replication.

FIGURE 7-14
Cartoon depicting our ignorance of the molecular machinery at the replication fork, which has been discreetly covered by a fig leaf.

This absurdly simple insight, which had eluded me for several years, proved a major turning point in the renaissance of our discoveries in replication.

Huge chromosomes are very fragile, and the mechanical forces of flow and mixing used during isolation are violent enough to reduce the DNA to a heterogeneous collection of damaged fragments. The chains have gaps, breaks in one strand or both, and ends that are uneven and frayed. In short, we were violating a basic tenet of enzymology by giving our relatively "clean" enzyme a very "dirty" substrate.

When in 1971 I finally recognized the futility of searching for replication enzymes in bacterial DNA, let alone animal DNA, I also realized that for five years we had been ignoring a proper DNA substrate, the chromosome of the tiny bacteriophages. I recalled belatedly the virtues of the intact, clean phage chromosome that four years earlier had served us in demonstrating the synthesis of infectious DNA by DNA polymerase I. As small, single-stranded circles, we could actually see them in the electron microscope and verify that in a purified sample they were intact, homogeneous, and uncontaminated by DNA of the bacterial host. We also knew that, immediately upon entering the cell, the phage DNA is converted by cellular enzymes to a double-stranded circle, an event we could easily assess by several methods. Probing how a new phage circle is started and completed might illuminate the intricate enzymatic machinery the cell uses to replicate its own chromosome.

We could use the chromosome of either of the two classes of small phages: φX174, whose DNA is packed into a roundish head, or the long, filamentous M13, whose DNA is stretched out. In both cases, immediately upon injection of the viral DNA, cellular enzymes start a new DNA chain to convert the single-stranded circle to the duplex form. Since DNA polymerases are not able to start a chain, something else must do so. Here was our chance to find out what this something else was. Later events proved that I was lucky in choosing M13 over φX174 to begin these experiments. I was prompted to do so by unusual ways in which the cell's membranes are involved in the disassembly of M13 at the cell surface and the assembly of new phage particles—aspects of viral infection that had intrigued me during my previous sabbatical year in England.

Having decided to work on M13, I met with my research group the

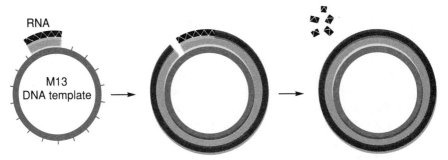

FIGURE 7-15
RNA primes the start of a DNA chain and is removed with completion of the chain.

next day and was able to persuade them of the importance of switching all our efforts to M13. Immediately, we started reading the M13 literature, and within a week all my students and postdoctoral fellows had dropped their previous projects and were working on various stages in the life cycle of M13. I have sometimes felt wistful about the boldness of that move and the exciting events in the weeks that followed.

My preoccupation with the initial event in M13 replication enabled me to connect three otherwise unrelated facts and arrive at an idea as to how a DNA chain might get started. As background, recall that nucleotides of RNA are distinguished from those of DNA by having an oxygen atom at the indicated position on the ribose sugar ring (Fig. 7-11), hence the name *ribo*nucleic acid for RNA and *deoxy*-*ribo*nucleic acid for DNA.

> *Fact one:* RNA polymerase, the enzyme that copies (transcribes) the genetic message in a DNA template into RNA language, differs from DNA polymerase in a fundamental way. RNA polymerase can start chains (Fig. 7-15, step I), whereas DNA polymerase cannot.

> *Fact two:* Although DNA polymerase routinely excludes ribonucleotides in assembling a DNA chain, it does accept an RNA chain end matched to a DNA template as a primer to be elongated for DNA synthesis (Fig. 7-15, step II).

> *Fact three:* DNA polymerase I has an editing function which can remove something foreign from the start of a DNA chain and replace it with proper DNA (Fig. 7-15, step III).

Could this be the scenario? (1) RNA polymerase makes a short piece of RNA on single-stranded M13. (2) DNA polymerase uses the RNA to start a DNA chain. (3) Upon completing the copying of the available template, the enzyme, in its editing function, recognizes the piece of RNA as foreign, erases it, and synthesizes proper DNA in its place (Fig. 7-15; see also Fig. 7-9). Michael Chamberlin happened to be visiting the laboratory the day I had this idea. An outstanding graduate student with Paul Berg in our first class at Stanford, Mike had joined the faculty of the Biochemistry Department of the University of California at Berkeley and quickly became one of the leading authorities on RNA polymerase. Mike agreed that RNA initiation of a DNA chain would make good sense. At dinner that evening, Sylvy and my son Roger were just as enthusiastic. The next day I discussed the idea with Doug Brutlag, my graduate student. There was an easy way to test it, and he was eager.

We could use the drug rifampicin, an antibiotic that blocks the growth of *E. coli* and related bacteria because it inhibits RNA polymerase. If synthesis of RNA by RNA polymerase were an essential first step in the replication of invading M13 DNA, then rifampicin should block it. We had both the drug and the assay. Within twenty-four hours the results were in. Rifampicin completely prevented the conversion of the single-stranded M13 circle to the duplex form! Might this be an effect of rifampicin on another target? We obtained a mutant strain of *E. coli* resistant to rifampicin because its RNA polymerase was not affected by it. When we infected this mutant with M13, the first stage of replication was unperturbed by rifampicin. "It now seems plausible," we wrote in a publication submitted to the *Proceedings of the National Academy of Sciences* in September 1971, "that RNA polymerase has some direct role in the initiation of DNA replication, perhaps by forming a primer RNA that serves for covalent attachment of the deoxyribonucleotide that starts the new DNA chain."

In the next few months we obtained evidence from extracts of *E. coli* that established what we had inferred from observations with intact cells. "RNA Synthesis Initiates In Vitro Conversion of M13 DNA to Its Replicative Form" was the title of our next publication. With the M13 viral circle as template, synthesis of a short length of RNA by RNA polymerase was extended by DNA synthesis around the circle.

There was one discordant note. Rifampicin did not affect the conversion of the rather similar single-stranded circle of phage φX174 to its duplex form either in vivo or in vitro. Nor did it interrupt the ongoing replication of the *E. coli* chromosome, despite the repeated initiations of DNA strands presumed to be occurring at the growing fork. We had to conclude either that RNA priming was of limited significance or, as we still hoped, that it was a mechanism of wide significance and that a novel mode of RNA synthesis, independent of RNA polymerase, was responsible for the priming of φX174 and the Okazaki fragments at replication forks. As we probed the striking difference between initiations on M13 and φX174 circles, it became clear that in M13's evolution, the virus developed a segment in its chromosome to exploit RNA polymerase, whereas the φX174 virus came to rely instead on initiation proteins that the host cell uses for replicating its own chromosome.

Wheels within Wheels of the Replicating Machine

I never imagined that starting and completing a short circle of phage DNA could be so complicated. Four years earlier, in 1967, to wide acclaim, we thought we had been able to accomplish it in the test tube with just two enzymes: a polymerase to start and extend a chain and a ligase to join the ends. In 1988, after seventeen years of working on an operation the cell completes in a few seconds, we are still grappling with its awesome complexity. Seven different proteins (possessing more than twenty discrete components) are needed to start the chain and a comparable number of others are employed to complete it. It may sound like bad news that such a large fraction of a scientific lifetime has been spent on a small bit of the life cycle of an obscure bacterial virus. But the good news is that probing with the tiny, manipulable viral DNA has illuminated the cellular operations on the massive, inaccessible host chromosome. This is because the initiating viral event appropriates the very machinery the cell uses to make its own DNA.

Expressing the juice of a bacterium to get at the numerous replication proteins, intact and freed from the chromosome and membranous coats to which they may have been bound in the cell, is tricky. In 1972 many months of empiric effort were devoted to finding the right strain, one whose cells could be disrupted under the gentlest

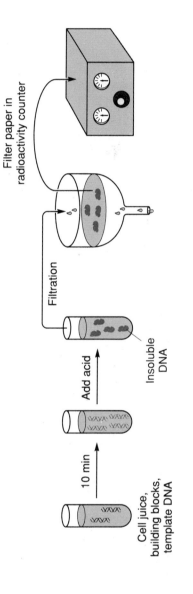

FIGURE 7-16

Assay of DNA polymerase activity depends on the incorporation of acid-soluble building blocks into DNA, which is insoluble in acid.

conditions to yield a bountiful harvest of the proteins we were seeking. After many hundreds of preparations, the procedure for collecting massive amounts of a microbe and gently opening the astronomical numbers of them is still far from routine and requires meticulous attention to dozens of details.

Typically, the cells are grown in a 300-liter (75-gallon) tank at 37°C in a medium of glucose, salts, and yeast extract, continually stirred and aerated. After about six hours, with the cells still doubling every thirty minutes, the culture is sedimented rapidly in a centrifuge. About two pounds of bacterial cell paste are collected from a 200-liter culture. The paste, suspended as a thick soup, is distributed into "Seal-a-Meal" plastic bags and quick-frozen in liquid nitrogen (−196°C); these *E. coli* "pancakes" can be stored for many months in a deep freeze.

We thaw the packet of frozen cells slowly, keeping the temperature near freezing, and then add lysozyme, an enzyme that disrupts (lyses) the membranous sac enveloping a cell. (Lysozyme, an antibiotic substance first discovered in tears by Alexander Fleming, inspired the later discoveries of penicillin and its action.) We then centrifuge the lysed cells to remove the sedimented membranous debris and DNA and to collect the precious proteinaceous juice.

A prime requirement in resolving and purifying enzymes, especially when several are involved in a multistep process, is a quick assay. (An assay of twenty samples in two hours encourages the trial of many alternative purification procedures.) In the assay of φX174 DNA replication, we measured the synthesis of a new DNA chain on a phage circle by the incorporation of radioactively labeled nucleotides into DNA. Upon acidifying a solution, the DNA becomes particulate, whereas the free nucleotides remain in solution. The radioactivity in particulate DNA collected on a filter paper disc can be measured in a counter (Fig. 7-16).

What strategy guides the separation and isolation of each of the numerous proteins in a cell extract responsible for a complex series of reactions? One approach, called *complementation*, requires genetic information about the proteins; the other way, called *resolution*, is purely biochemical.

For complementation to be used, a mutant deficient in one step in the process must be in hand. An obvious limitation of this approach is the need to have mutants for each of the many genes for replication, some of which are still unknown. Another problem is that mutants, as a consequence of their illness, usually have secondary deficiencies in other replication proteins which normally depend on the stabilizing interactions with the unmutated proteins; as a result, the mutant extract may be responsive to complementation by proteins in addition to the one being pursued. Finally, should isolation of the protein be successful to the point of having it pure, too little is learned of this "pearl" when added to the "swinish" disorder of a crude extract.

A special way is needed to obtain an altered replication gene, inasmuch as such a defect would likely make it impossible for the cell to multiply. As mentioned earlier, a selection is made for a mutant which encodes an enfeebled protein that can function at 30°C but fails at 42°C. Among the known temperature-sensitive mutants for DNA replication, one called *dnaG* permits the normal φX174 phage infection at the low temperature but not at the high; similarly, extracts from *dnaG* mutant cells functioned only at the lower temperature in our assay measuring the start and extension of a new chain on a φX174 DNA circle. One more fact: The dnaG protein from normal *E. coli* cells, whether in a crude or purified form, when added to an extract of the mutant cells, enabled it to function at 42°C. Thus, the mutant cell extract, inert at the restrictive (high) temperature, needs only normal (heat-resistant) dnaG protein to complement its deficiency and thereby provides a basis for detection and quantitative assay of dnaG protein.

The other "resolution" approach to the separation and isolation of proteins can be illustrated by what we have been doing routinely for many years. Ammonium sulfate salt is added to the cell extract. The amount, determined by trial and error for each purification procedure, causes some of the proteins to aggregate and become particulate while others remain in solution. Sedimentation in the centrifuge separates the insoluble fraction (later redissolved) from the soluble (Fig. 7-17).

In pursuing the replication activity present in the crude extract, we failed to find it in either the sediment (Fraction I) or in the soluble matter (Fraction II). We did recover the activity when both fractions were combined. Clearly, at least two proteins were essential and had been separated in the two fractions. (By using complementation assays, as described above, we found the dnaG protein only in the sedimented fraction.) In further pursuit of the activities in Fractions I and II, we tried a variety of adsorbents. Upon passing

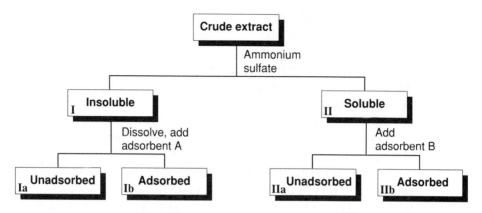

FIGURE 7-17
Early steps in fractionating (separating and purifying) the numerous enzymes required in replication of the single-stranded φX174 circle.

Fraction I through a column of adsorbent A, the activity appeared
neither in the unadsorbed passthrough fluid (Fraction Ia), nor in the
strong salt solution that releases adsorbed proteins (Fraction Ib) but
was fully recovered when Ia and Ib were combined. Similarly, Fraction II could be further fractionated into IIa and IIb. Now it seemed
we had at least four distinctive proteins (the activities in Fractions
Ia, Ib, IIa, and IIb) (Fig. 7-17) that were needed for the start and rapid
elongation of a DNA chain on the φX174 template.

After a ten-year drought of discoveries of new enzymes, they were
now coming in a torrent, and in 1972 a bright and boisterous group
was there to collect and sort them. Leading the team were Randy
Schekman and Bill Wickner. Randy, a beginning graduate student,
was already advanced in techniques and had a taste for exploring
mechanisms of DNA replication. Bill, trained in medicine at Harvard, had research experience in lipid biochemistry under the
guidance of Eugene Kennedy and knew the power of enzyme
purification. Bill's overflowing enthusiasm got us into some trouble.
Daily telephone bulletins to his sister-in-law, Sue Wickner, working
as a graduate student with Jerard Hurwitz at Albert Einstein Medical
School, encouraged them to adopt the same course we were taking
and to make it a race.

I did not object to Bill's calls because I am uncomfortable with
secrecy in science. Uninhibited daily exchange of information is

essential among members of a research group and, with care to avoid conclusions from half-baked results, should be shared promptly with colleagues in other groups, visitors, and interested scientists elsewhere. Occasionally, the use of such unpublished information, without proper restraint and discretion, taints the priority of an important discovery. (Inadequate attribution to published discoveries is even more common.) Such annoyances can be rankling but are outweighed by the benefit and pleasure of sharing the excitement and puzzlement over fresh data and their possible significance.

Yet I am also uncomfortable with the emphasis on competition in science. It is no fun for me to strive to do something that is likely to be done just as decisively by someone else at about the same time. Unless it is urgent to clear away an obstruction in the path to a major objective, I would rather work on one of the many problems around the core of my interests and competence, one that is not likely to be solved soon by others. Jerry Hurwitz feels differently about progress in science. His broad interests and ambitions, combined with his experimental skill, knowledge, and accomplishments, have left him few territorial inhibitions. Like so many others, he enjoys the jostling and noisy excitement of a scientific race.

After we began to purify the components responsible for the replication activity, we could separate them into two groups, one that primed the start of a chain and the other that extended it. Then, as we tried to purify each of these fractions, they splintered into many separate components. The joy of uncovering the trails to so many novel proteins soon gave way to the discouragement of being unable to track down any one of them. We judged there might be as many as eight different proteins needed to make the tiny bit of RNA that primed the synthesis of a DNA chain. What were all these proteins and what was each doing?

Purifying each of these proteins and obtaining them in useful amounts was a formidable task. (Fifteen years later, one of them is still in an unpurified state.) To begin with, they are rather scarce in the cell, each about 1 part in 10,000. A successful isolation of one of them, based on three years of exploratory work, yielded only 3 milligrams from 3 kilograms of bacterial paste after a solid month's effort. Such precious, hard-won materials were husbanded in a deep-freeze vault and kept out of reach of new students, with the result that progress was rather slow. The assays were also troublesome. In the crude state in which we had these proteins, traces of a given activity were found among several fractions, making the measurements of the enriched fractions less decisive and reliable. Some of the proteins were unstable and some stuck to one another. It also mattered in what relative amounts and in which order the fractions were mixed with each other, and with DNA and the building blocks, when assembling the recipe of the nearly thirty ingredients for replication.

I will be forever indebted to the sewage of New Haven and the phage G4 that Nigel Godson of Yale University isolated from it. Searches in sewage had uncovered some of the other phages that prey on the bacterial denizens of the human gut. The new isolate, phage G4, bore many resemblances to φX174 and prompted us to compare the behavior of its small circular DNA as a template for replication. What a lovely surprise! Replication of G4 DNA by the crude cell extracts was more vigorous than that of φX174 DNA and, most important, far less demanding. Only three, rather than eight, fractions were needed. We were now able to purify them to the point of understanding each of their functions.

We were happy to discover that one of the fractions could be purified very easily because it withstood boiling temperature, a treatment that coagulates and inactivates more than 95 percent of cellular proteins. This protein binds and coats all available single-stranded DNA, except in certain places where the matching of bases permits the DNA to fold back on itself in a hairpin-like double helix (Fig. 7-18). This coating of all but 1 percent of the DNA by the single-strand binding protein (SSB) directs the action of the next protein.

A second fraction proved to be the protein that complements the deficiency of dnaG mutants. This protein has evolved the ability to recognize the uncoated hairpin-like segment unique to G4 DNA and synthesize a stretch of RNA to match it. Because this RNA serves as a primer for DNA polymerase to start the DNA chain, we renamed the dnaG protein *primase*. (The DNA of φX174, for lack of the particular G4 sequence, requires a rather elaborate set of proteins to enable primase to do its work.)

The third fraction was DNA polymerase. As will be described in Chapter 8, this highly complex form of DNA polymerase III is equipped with many auxiliary units that enable this enzyme to replicate enormous lengths of DNA with astonishing speed and accuracy.

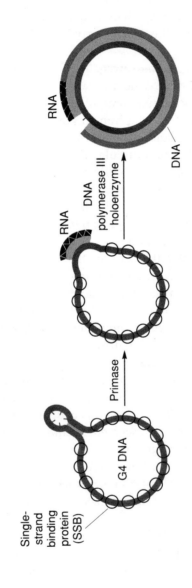

FIGURE 7-18
Stages in the conversion of the single-stranded circle of phage G4 to the duplex form.

Replication of G4 DNA provided us with our best assay for purifying this super polymerase of *E. coli*, which we now called DNA polymerase III holoenzyme.

Are you *still* working on DNA replication? How can I convey to the curious and sympathetic people who ask me this question that the perimeter of ignorance about this subject is far greater than when I started over thirty years ago and that the search for solutions is even more exciting?

The awesome and complex assemblies of proteins that start and extend DNA chains will consume the efforts of scientists for decades to come. Even when we identify each of the many parts of these machines, locate and clone their genes, amplify these proteins by genetic engineering, and then put an operating humpty-dumpty together, we will still be far short of what we need to know about the replication process. We will need to know the arrangement of these components down to atomic detail in order to explain the extraordinary speed and precision of making a DNA chain. Further, we will want to know how the replication genes are used to make the corresponding proteins in precisely the right amounts and ratios at just the time they are needed. Beyond the catalytic activities of these assemblies and the factors that fine-tune their operations, we need to be aware of their social relations in the cell—their attachments to other protein units, to membranes, and to the skeletal framework—if we are to understand how the replication of a chromosome is started and completed and how the two emergent chromosomes are correctly partitioned between the two daughter cells.

Chapter *8*

Frontiers in Replication

After having discovered a new entity and given it a name, scientists often slump back, exhausted by the effort, and leave for another wave of investigation the problem of understanding its composition and the mechanism of its operation. So it has been with the DNA replication machinery, just as it was with the transcriptional apparatus for expressing DNA information in the language of RNA and the translational devices for proteins.

The large, complex protein assembly that primes the start of a DNA chain we named a *primosome*. This molecular machine operates at the replication fork to lay down the priming starts of the "lagging" chain of the double helix replica. We have been able to examine some of its intricate structure and mechanics but have also been frustrated by its great fragility.

As for the multicomponent enzyme that builds the DNA chain, our awareness of its size and fragility prompted our naming the intact entity *DNA polymerase III holoenzyme*. This sewing machine, which assembles the DNA chains with extraordinary speed and fidelity, depends on more than twenty components. In size and intricacy, it rivals the ribosomal machines that make protein chains.

Encompassing the primosome and polymerase holoenzyme, we imagine a super polymerase (a *replisome*) that also includes auxiliary replication proteins, such as helicases, that unzipper the parental duplex to advance the replication fork. It would be aesthetically pleasing were a smoothly coordinated operation by the replisome to achieve virtually concurrent synthesis of both nascent strands rather than have the jerky, spasmodic dynamics of the uncoupled components. This prize will come with the discovery, disassembly, and reassembly of the wheels within wheels of these elaborate and awesome machines.

With at least some understanding of how DNA chains are made, one of the most fundamental questions remains. What is the switching process that turns the whole cycle of chromosome replication on or off? What is the mechanism responsible, on the one hand, for starting replication in embryonic or malignant cells that divide frequently and, on the other hand, for keeping it in check in cells that achieve the quiescence of adulthood?

Viruses, because of their unrestrained multiplication, are not suitable models for discovering the mechanisms that regulate the initiation of a cycle of chromosome replication. Biochemical studies with an intact bacterial chromosome, let alone an animal chromosome, have been beyond reach. However, an alternative approach became possible with the advent of recombinant DNA technology and plasmid engineering.

Plasmids are tiny duplex circles of DNA which occur naturally in bacteria. The size of small bacteriophages, they multiply separately from the bacterial chromosome. They may possess genes for proteins that destroy antibiotics, neutralize drugs, cleave DNA, and enable the plasmid to be transferred from one cell to another.

By excising the starting site of replication from an *E. coli* chromosome and inserting it into a plasmid, genetic engineering can create a minichromosome which behaves like the bacterial chromosome. It thus becomes a proper substrate for revealing the proteins that operate the switch initiating chromosome replication and control the growth and reproduction of the cell.

A more remote frontier is the final phase of replication, when the newly replicated pair of chromosomes must be partitioned correctly between the two daughter cells. The problem is especially complex in cells with many pairs of chromosomes. In the process of cell division, the membranous envelope of the cell is plausibly a participant in chromosome sorting between daughter cells, but direct evidence is still lacking. After a number of unsuccessful probings of the phospholipid fabric of the membrane, we may finally, after many years, be coming closer to a chemical role for a membrane component in replication. The use of tiny engineered plasmids as models may prove helpful. Many plasmids possess special means to ensure their efficient distribution between dividing cells. Thus far, our knowledge of these devices is based almost entirely on genetic evidence, and novel assays need to be developed to explore their biochemical nature.

Even though my work on DNA replication has been exclusively with the apparatus of *E. coli*, the evidence is compelling that basic biochemistry, including the synthesis of macromolecules (DNA, RNA, and proteins), is conserved throughout evolution. With regard to DNA replication, the enzymes and their actions discovered in *E. coli* during the past thirty years have thus far been prototypical for those found in subsequent studies of eukaryotic (plant and animal) cells. Intensive biochemical studies of eukaryotic systems can now be pursued widely and assiduously. Beyond the aesthetic appreciation of how Nature replicates its threads of life, there is the expectation that this newly acquired knowledge of the basics of replication will also help humans avoid some of the tragic aberrations of genetic diseases, cancer, and aging.

The Locomotive That Starts DNA Chains

However crude the analogy, the image of a locomotive was helpful for a time in explaining the molecular operations of the primosome, the protein assembly that starts a DNA chain. We were baffled at first in finding, let alone fitting, the components of the eight-protein jigsaw puzzle that describes initiation of replication of a phage φX174 circle. Only after we had purified two key protein pieces and discovered their functions by working on the simpler G4 phage system could we profitably return to the φX174 problem.

One of these proteins, we learned, coats the single-stranded DNA. About 200 molecules line up along the DNA, except at isolated hairpin-like duplexes. The other protein, the dnaG protein, renamed primase, seeks out one such hairpin region to lay down the RNA transcript which primes the DNA chain. Between these initial and final actions, primase and the other six proteins assemble and propel the locomotive. This structure is large, complex, and organized enough to resemble a ribosome, and so was named a primosome. Of

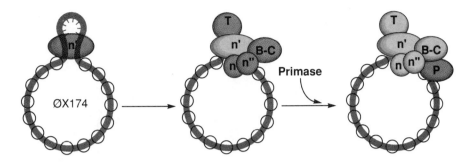

FIGURE 8-1
Stages in the assembly of the multiprotein primosome on the phage
φX174 circle.

these proteins, two are products of the *dnaB* and *dnaC* genes, mutants of which are known to be defective in replication; six
molecules of the dnaB proteins, in doughnut array, are matched by
six dnaC protein molecules. Another protein, named i, has only in
recent months been connected with its genetic origin, a replication
gene known as *dnaT*. The genetic sources are still unknown for the
remaining three proteins, which we isolated by fractionation procedures and gave the trivial names n, n′, and n″.

How is the primosome assembled and how does it work? We
identified a particular region of the φX174 DNA circle, about 40 of
the 5,386 nucleotides, which doubles up into a hairpin. Protein n′ is
excluded by a coating of the single-strand binding protein (SSB)
from binding the DNA molecule, except at this hairpin structure.
There n′ attracts the sequential deposition of the other proteins (Fig.
8-1). The twenty-part machine is now ready to operate.

Protein n′, powered by ATP energy, serves as an engine and also
has a cowcatcher to remove SSB covering the DNA in its path. The
dnaB protein is the engineer. It uses ATP energy to locate or shape a
section of DNA track upon which primase will find it possible to lay
down a short stretch of RNA, which will then attract DNA polymerase to start a DNA chain.

When we examined the behavior and fate of these proteins assembled into a primosome on the DNA circle, we were astonished by
the virtuosity of Nature's chemistry, embarrassed by our failure to
anticipate it, and then frustrated by its elusiveness.

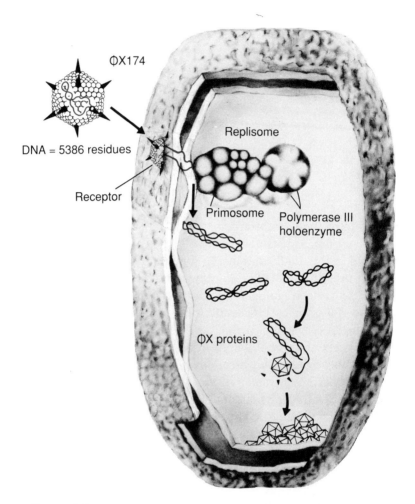

ΦX174

DNA = 5386 residues

Receptor

Replisome

Primosome

Polymerase III
holoenzyme

ΦX proteins

FIGURE 8-2
Life cycle of phage φX174. The duplex (formed from the single-
stranded circle), with the primosome attached, becomes a stamping
machine for making many copies of the duplex. Single-stranded circles generated from a duplex are enveloped in protein coats and released as infectious particles.

To examine the primosome, we added the several proteins needed to initiate replication, having radioactively tagged each to trace their fates. We found that they assembled in a regular ratio and became bound rather firmly to the DNA. One of the major surprises was that the primosome, once formed, moved rapidly around the DNA circle and could generate many primers on it. This at once explained a long-standing observation that the start of replication of infecting φX174 DNA was not at a unique place, as with M13 or G4 DNAs, but rather at many places around the circle. Movement along the DNA was in only one direction of the backbone strand, designated 5→3. The polarity of this movement is likely to be important in advancing the replication fork of a duplex chromosome.

Another unexpected primosome property was the persistence of its attachment to the φX174 DNA circle long after its priming job was done, and even after the duplex circle had been completed by polymerase and sealed by ligase. Why did the primosome remain attached to the DNA? Now it seemed obvious. Once formed, the primosome would not be used just once and discarded. Rather, fixed to the duplex circle, it would be used again and again as part of a stamping machine in the many repetitions of multiplying the phage DNA molecule. From this machine come the hundreds of phage DNA circles, each to be enveloped in a protein coat and released as an infectious virus particle (Fig. 8-2).

Probing into how duplex circles were multiplied, we discovered that a protein encoded by the phage nicks one of the strands at a specific place and attaches itself to one chain end at the break to start a process ("rolling-circle" replication) that generates a stream of phage DNA circles. A fascinating revelation was that, for this to happen, an E. coli protein (called a helicase) uses ATP energy to unzipper the duplex to expose the template for the advancing fork of replication.

We could see the primosome straddling the φX174 DNA circle in the electron microscope, but this picture with its intriguing loops (Fig. 8-3) did not answer the most basic questions. How was it fashioned to bind the DNA and move on it in a unique direction? How was ATP energy used in its movement, and how much was consumed in its circuits? How did the primosome increase by 20-fold the speed with which another enzyme later split a particular bond (one among the 5,386 in the circle) in the next stage of repli-

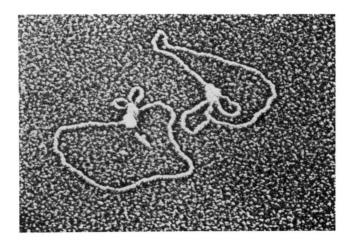

FIGURE 8-3
Electron microscopic view of the primosome attached to
DNA duplexes formed from the single-stranded circle of
phage DNA. (Courtesy of Dr. Jack Griffith.)

cation? We were poised and eager to tackle these and many like
questions when the monster we always fear showed its fangs.

It struck Bob Low, a postdoctoral fellow who had expanded the
primosome studies started by another fellow, Ken-ichi Arai. Upon
returning in early 1981 from an extended Christmas holiday, Bob
was unable to repeat the isolation of an intact, active primosome.
During his absence, some of the partially purified protein preparations had been used up by others and replaced. Were these new
samples lacking an unknown factor supplied in the previous preparations? Or were novel impurities present which interfered with the
delicate associations of the numerous proteins needed to create and
preserve the primosome? Perhaps a subtle change had been introduced into the reagents or procedures used for isolating the
primosome.

After more than a year of trying hard to recapture the primosome,
it was time to retreat. Bob was exhausted and depressed by the effort,
and other projects looked brighter. Then three years later, we decided to pursue the elusive primosome once again. This time, new
faces joined the chase, reinforced with more plentiful, genetically
engineered stocks of proteins and, most important, with a fresh
sense of mission.

The Super Sewing Machine

Imagine a sewing machine that can survey a mountainous, coiled-up tangle of ribbon and instantly locate a tiny unique spot at which to start copying its pattern. It selects each stitch from one of four colors and sews 1,000 stitches per second to produce in a new ribbon a flawless copy of the old. By recognizing and removing any incorrect stitch, the machine reduces mistakes to one in ten million, sewing without interruption for forty minutes until the perfect replica of the 4-million-unit length of ribbon has been completed. This sewing-machine enzyme, with all its working parts, we named the DNA polymerase III holoenzyme of *E. coli*.

Only a fraction of the capacities of the polymerase holoenzyme are included in this gross description. Very likely, the enzyme's replicative activities are regulated and its position in the cell oriented by "antennae" that respond to various signals in the environment and "social" contacts with other enzymes. These additional features, still unexplored, are needed for integration with other replicative enzymes, as well as those responsible for the simultaneous repair and rearrangement of the chromosome and the concurrent transcription and expression of its genetic information.

We became aware of the awesome properties of this machine in 1972 when we were trying to do an apparently simple job: the rapid conversion of a single-stranded phage DNA circle to the duplex form. With an RNA primer put in place by the primosome to start the chain, none of our purified polymerases, neither I, II, nor III, could synthesize a DNA copy at a significant rate. We looked for and found such a polymerase activity in crude cell extracts. It proved to be a highly complex form of polymerase III, containing ten distinctive proteins, with two or more copies of each.

We have grappled with this most difficult enzyme for fifteen years and are finally close to having a homogeneous preparation. We call it "holoenzyme" to distinguish it from forms that lack one or more of the components. Many attempts have been made to separate the holoenzyme into its subunits and then reconstitute them into a fully functional holoenzyme. This work in my laboratory and those of former associates (for example, Charles McHenry, now at the University of Colorado in Denver, and Robert Bambara at the University of Rochester) has been very difficult because of the fragility of the enzyme and its paucity in the cell. With only 10 to 20 copies in an *E. coli* cell, compared with thousands of most enzymes, we could get only a few micrograms of the purified enzyme from several kilograms of cell paste. Like some rare first editions, these precious samples were admired rather than used, and relatively little progress was made—until genetic engineering came on stage.

There is a widespread and justified impression that recombinant DNA technology is performing miracles for medicine, agriculture, and industry, not to mention Wall Street. Development of insulin, interferons, interleukins, and vaccines has had an extraordinary impact on diagnosis and treatment of disease. But to me, those achievements are dwarfed by what genetic engineering has done for biomedical science. By cloning the gene for a given subunit of the polymerase holoenzyme, amplifying its protein products many times beyond its normal level, and isolating the protein in an active form, we have obtained most of the subunits in quantities sufficient to examine their individual capacities. We can constitute these subunits into functional groupings and assemble a near-complete holoenzyme with a catalytic efficiency and a fidelity of replication approaching those of the enzyme in an intact cell (Fig. 8-4). Inevitably, as we reexamine this reconstituted assembly and devise more exacting assays, it will be found wanting in some particulars and, as with other man-made constitutions, will need amendments.

As an example of this exercise in assembling the individual subunits, I will consider two of them, the principal elements of the functional core of the enzyme. The one called alpha is the polymerase, the sewing unit; the role of the other subunit, called epsilon, is to remove nucleotide units from the growing end of a DNA chain. The gene for epsilon is the very one discovered ten years earlier and called *mutD* because defects in this gene raise the level of spontaneous mutations in *E. coli* 100,000-fold. Clearly, a defective epsilon subunit fails in its job of proofreading during replication, thus leading to the retention of numerous mutations. What we found most gratifying was that, by mixing together the alpha and epsilon subunits, we produced a complex with a potency far greater than the sum of the two parts: the polymerase activity doubled over that of alpha alone, and the proofreading capacity of epsilon increased 50-fold. We could show that the alpha subunit improved the proofreading efficiency by directing the epsilon subunit to its site of action— the growing end of the chain; epsilon, in turn, also improved the polymerizing efficiency of alpha.

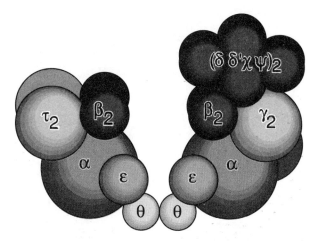

FIGURE 8-4
The ten distinctive subunits of DNA polymerase III holoenzyme arranged as a twin assembly of the polymerase entity
(subunit alpha, α) with the auxiliary subunits asymmetrically
placed.

The function of the other subunits is to provide a clamp which
holds the core to the template and moves with it in the course of
replication. One, called tau, raises the efficiency with which the
holoenzyme finds and binds the primer end of the DNA and then
clamps the polymerase to the template to prevent it from dissociat-
ing each time it inserts a nucleotide unit and moves up a notch to
add the next. Another group of five (gamma, delta, delta prime, chi,
and psi) also furnish a clamp, perhaps a more finely regulated device
to govern attachment of the polymerase to the template.

Under the Fig Leaf at the Replication Fork

Based on the insights we gained from the replication of the small
phages as to how DNA chains are started, elongated, and completed,
we could finally peek under the edges of the "fig leaf" covering the
replication fork (Fig. 7-14) and begin to see some of the working
parts of the molecular machinery and imagine how they might drive
the replication of the duplex chromosome.

Helicases, the enzymes that use ATP energy to unzipper (unwind)
the duplex in advance of the fork, expose single-stranded DNA to

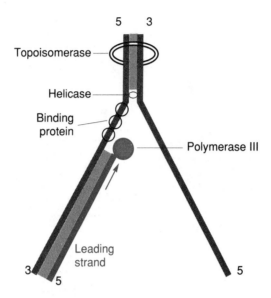

FIGURE 8-5
Current view of the enzymes that open the duplex to expose the template for *continuous* synthesis of the leading
strand in replication.

serve as templates. The untwisting required for unwinding duplex
DNA is catalyzed by enzymes called topoisomerases, which provide
a swiveling action. Single-strand binding protein covers the exposed
single strand (Fig. 8-5), protecting it from destruction by ubiquitous
chain-cleaving enzymes (nucleases) and putting the chain in a
configuration optimal for its template function.

The multisubunit polymerase builds the DNA chain along one
template strand rapidly and continuously. This becomes the *leading*
strand in advancing movement of the replication fork. On the other
exposed single strand, now also covered by binding protein, repeated initiations by the primosome generate the *lagging* strand,
discontinuously (Fig. 8-6). The primosome lays down a short RNA
primer and then moves on the template strand in the direction of
fork movement, thereby positioning itself at the fork to destabilize
and unzipper it further. Each primer is extended by DNA synthesis.
This nascent DNA (the Okazaki fragment) fills in the available template until it reaches the previous fragment. DNA polymerase I, performing a function that DNA polymerase III cannot, recognizes the
primer RNA backbone on the previous fragment as foreign, removes

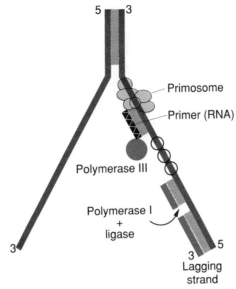

FIGURE 8-6
Current view of the enzymes that initiate the RNA-primed
starts in the *discontinuous* synthesis of the lagged strand;
the foreign RNA is later removed, gaps are filled, and nicks
are sealed.

it, and fills the resulting gap. Finally, ligase seals the joint between
the two fragments.

Replication of a duplex chromosome is therefore semidiscontinuous: continuous on one side, discontinuous on the other. While correct overall, the operation, in molecular detail, seems much too jerky
and clumsy for the elegance we expect of Nature's chemistry. We
imagine a more efficient device that embraces the polymerase, the
primosome, and helicases—in essence a super entity, referred to
earlier as a replisome, that might be able to carry out the replication
of both strands *concurrently* rather than spasmodically.

The cell is not just a bag of loose enzymes. Rather, they are joined
in family units in specific locations to carry out their particular
functions. Because the bonds that link them are often too tenuous to
withstand the disruptive forces necessary to open a cell, we have yet
to take the replisome "alive." Beyond the plausibility of its existence, our fragmentary data are consistent with such a structure. The
holoenzyme, as we isolate or assemble it, contains two rather than
one of each of the subunits. Thus, a twin-assembly of polymerase

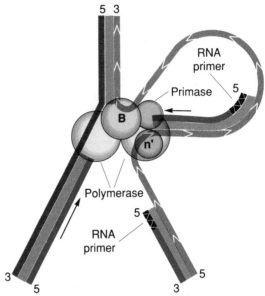

FIGURE 8-7

Hypothetical view of the replisome as a fixed machine through which a loop of DNA is pulled to provide for essentially concurrent replication of both strands. The direction of the DNA template in the two arms of the loop, as it is pulled through the replisome, translocates n' protein on the DNA in the direction of chain elongation and the B protein in the opposite direction.

units in the holoenzyme may be juxtaposable to perform concurrent replication of both strands and coordinate the priming action of the primosome with the DNA synthetic activity.

Two dilemmas confront the proposal of a locomotive replicating the two parental strands in a largely concurrent fashion. One is that the strands in a DNA duplex are oriented in opposite directions, whereas the direction of DNA chain growth by polymerase is invariably in one direction, as we have seen. This dilemma can be circumvented if a small portion of the lagging strand template is turned 180° by the formation of a loop (Fig. 8-7) to give both templates the same direction for a limited time. The DNA backbone, especially when it is single-stranded, is very flexible and can be bent back and forth with great ease.

The other dilemma became apparent when we examined the separate parts of the primosome "locomotive" and found that, while the

B protein and the primosome itself moved on the DNA track in the direction of progress of the replication fork, the n' protein and primase components moved on DNA in the other direction. Clearly, with key parts of the locomotive pulling in opposite directions, this metaphor would have to be abandoner. We must now revise our notions about the motions of the replisome and DNA relative to each other. Plausibly, the replisome is a stationary machine, perhaps fixed to the skeletal framework of the cell, with the DNA template drawn through it. To pull a loop of DNA through the replisome (Fig. 8-7), it can be seen that, relative to the proteins, the two arms of the loop move in opposite directions.

When we examine what has already been learned of the replication of DNA in a large variety of bacterial, fungal, plant, and animal cells, and of the viruses that infect them, we find that themes discovered in *E. coli* are played over and over with only minor variations. The generality of the basic biochemical patterns, as evident in the glucose metabolism of yeast and muscle, has now been seen again in the biosynthesis of the major macromolecules: DNA, RNA, and proteins. *E. coli*, the goose that has produced so many of these golden eggs of discovery, still retains its fertility. Through its long evolution, mechanisms have been highly refined that have also been adopted for the operations of the cells, of so-called "higher organisms."

The Switch That Initiates Replication of a Chromosome

What is the mechanism that controls replication? What is the biochemistry of the switch that turns a cycle of chromosome duplication on or off? These simple questions must concern and fascinate everyone who has ever thought about the nature of cell growth and its aberrations. The onset of chromosome DNA replication is the crucial event that distinguishes an embryonic cell from a mature one, a cancer cell from a normal one. Yet the chemistry of this event is a "black box." Virtually nothing is known of its contents in animal cells, and our ignorance about bacterial cells was, until recently, nearly as abysmal. I felt that our best route to solving this basic enigma was to expose and examine the control of replication in bacterial cells.

Chain growth - 40 min   Division - 20 min

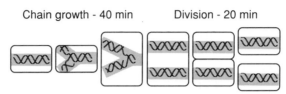

FIGURE 8-8
The cell cycle of E. *coli* is controlled by initiation of replication of its chromosome.

Bacteria grow and divide at different rates, depending largely on
culture conditions, despite the fact that the rate of growth of the
DNA chain is relatively inflexible. In most E. *coli* strains at 37°C, it
takes about 40 minutes to duplicate the chromosome (Fig. 8-8). The
time needed to build a wall (septum) inside the cell, to segregate
the duplicate chromosomes on either side of it, and then to separate
the two daughter cells is also fixed—it takes about 20 minutes. How
is this hour-long process accomplished in bacteria that can double
their mass and divide every 20 minutes? And why do the same
bacteria, under other nutritional conditions, take 200 minutes for a
process that could be accomplished in 60 minutes?

In rapidly growing cells, fresh initiations of the chromosome are
made on previously initiated chains that are not yet completed. New
pairs of chromosomes are being made even as the previous ones are
being segregated (Fig. 8-8). By contrast, in slowly growing cells that
divide every 200 minutes, a long delay follows the completion of cell
division. Thus, the major variable in replication is the decision
when to initiate the process. As in other biosynthetic processes
(such as synthesis of nucleotides and amino acids and their assembly into nucleic acids and proteins), the crucial decision is the initial
one, which commits the cell to embarking on a long and ener-
getically costly project.

Two major developments allowed us to begin to explore the
biochemical features of initiation of chromosome replication. First
was identifying a sequence of 245 base pairs as the unique origin
(oriC) of E. *coli*'s 4-million-base-pair chromosome (Fig. 8-9) and then
cloning this sequence in a plasmid, an engineered circle of DNA.
The second major development was our discovery of the enzyme
system responsible for replication in these plasmids.

Replication of oriC plasmids is subject to the same controls that

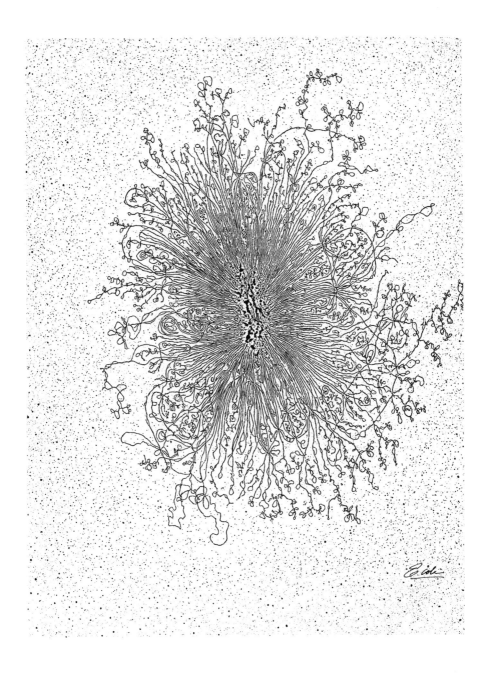

FIGURE 8-9
Electron microscopic view of the *E. coli* chromosome, an intact duplex
circle of four million pairs of nucleotide units. The origin of replication
occupies a region of 245 pairs, a length of DNA that would be barely visible on this scale. (Courtesy of Dr. Ruth Kavenoff.)

govern replication of the host chromosome: both depend on the protein product of the *dnaA* gene (known to be essential for starting a new round of replication), both are replicated bidirectionally from the initiation point at *oriC*, and both start replication at the very same time in the cell cycle.

When in 1977 Sei-ichi Yasuda and the late Y. Hirota in Japan succeeded in creating an *oriC* plasmid through genetic engineering, and Yasuda brought the plasmid to my laboratory, it seemed reasonable that we would soon find the enzyme systems for its replication, much as we had done for several phages of comparable size. This proved far more difficult than I had imagined. When Yasuda returned home two years later, there had been no discernible progress. Five people in my group spent an aggregate of ten years without any significant success. One of them was Bob Fuller, a graduate student. By early 1981 he had spent two frustrating years trying to prepare a cell extract that could replicate an *oriC* plasmid. He had learned how to detect and avert DNA syntheses not dependent on the *dnaA* gene and had improved the sensitivity of the assay 100-fold. But still there was no glimmer of activity. "Bob, it's time to switch to something else for your thesis—maybe one of the interesting primosome proteins." Classmates working on other projects in the department had already published papers and were presenting reports at national meetings. "Give me three more months," he answered. After considerable discussion, I agreed.

Bob had entered Yale intending to be a history major, but he dropped out after his freshman year. He traveled as a guitarist with a rock band for three years and then returned to finish with a degree in science and an impressive piece of undergraduate research. The three-year truancy made me uncertain about accepting him for graduate work, but his brightness and subsequent record prevailed. He has since done excellent research, won the most prestigious postdoctoral fellowships, and has just returned to Stanford to join the faculty of the Biochemistry Department.

One day in June of 1981, still within his three-month grace period, Bob (working with Jon Kaguni, a postdoctoral fellow) found the formula for a highly specific cell-free system for replication of *oriC* plasmids. It was more powerful than we had ever hoped for. The gates were now opened for us to discover the parts and operations of the mechanism that initiates the duplication of a chromosome.

Discovery of the cell-free system depended on two unlikely conditions. One was the inclusion of a high concentration of a large synthetic polymer, such as polyethylene glycol, which is used in the manufacture of paints, polishes, and cosmetics. Why? The polymer was there as a vestigial remnant of a complex recipe that promotes RNA and protein synthesis. We introduced it at a time when it seemed possible these other biosynthetic processes might be needed to start replication. Although these experiments led nowhere, the polymer was retained in subsequent trials. Such polymers are known to take up space in a solution and, by excluding other large molecules (such as DNA and enzymes) from their space, crowd them and make them more interactive, as we believe they are in a cell. We later found that the synthetic polymer also causes the clumping and effective removal of some inhibitory substances.

The other unlikely condition that led us to the discovery of the active system was doing something that everyone thought unreasonable. We subjected a totally inert cell extract to a refined salt fractionation. It was a maneuver that had worked for me thirty years earlier and which Jon Kaguni, after some coaxing, tried and did well. He added the salt (ammonium sulfate) in graded amounts to the extract and collected fractions of the protein precipitated by each of the salt additions. Each of these fractions was, in turn, dissolved and tested. One of them was active beyond our wildest dreams. The fractions produced by adding slightly less or more salt were completely inactive. Less of the salt failed to precipitate some of the numerous proteins needed to sustain the reaction; more salt brought down potent inhibitors, which blocked the reaction or destroyed the product.

With the soluble enzyme system (Fraction II, as we called it), we could begin to dissect out the many proteins of the "switch" that throws the chromosomal origin into the "on" position for replication. It pleased me, too, that this same fraction made it possible for several research groups to explore the replication of other plasmids and phages which had previously been beyond reach.

Even though partial genetic sequences of diverse creatures are reported daily, and even though the entire sequence of the four billion base pairs of the human genome will probably one day be known (a multibillion-dollar project), the origin of the *E. coli* chromosome (oriC), which is only 245 base pairs long, still stands

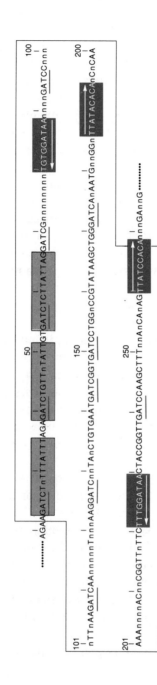

FIGURE 8-10
The sequence of nucleotide units that constitute the origin of the *E. coli* chromosome. This sequence is highly conserved in bacterial evolution.

out as a unique and impressive genetic sequence (Fig. 8-10)—for these reasons:

(1) *OriC* sequences in bacterial species remotely related to *E. coli* and separated by 200 million years of evolution are astonishingly alike, even though nearly all of the rest of their DNAs are widely divergent. Long stretches of highly conserved sequences in *oriC* are separated by "spacer" regions of fixed length but random sequence. Mutation of a single base pair (such as substitution of AT for GC) in a conserved region, but not in a spacer region, impairs replication; changing the length of the spacer by deletion or insertion of a base pair also has a deleterious effect.

(2) The sequence TTATCCACA, appearing in four places in conserved regions in *oriC*, has proved to be the specific binding site of the dnaA protein. These "dnaA boxes" are clearly essential for initiating replication.

(3) The sequence GATC, which would on a statistical basis be expected to appear once, is found 11 times in *oriC*. The adenine (A) of the GATC sequence is selected by a special enzyme in *E. coli* for attachment of a methyl (CH_3) group, whose effects on replication are now under intensive study.

(4) The sequence GATCTnTTnTTTTn (n being any nucleotide), repeated three times in *oriC*, has a crucial function in a stage following the binding of dnaA protein to its "boxes."

The formidable array of conserved sequences, separated by precisely placed and measured spacers, may endow *oriC* with the capacity to assume a unique three-dimensional shape. We can also expect that recognition and use of this complex structure to initiate replication must depend on an assembly of proteins of comparable complexity, equally conserved over long spans of evolution. What are these proteins?

Over a period of five years, we identified 13 proteins as participants in replication of *oriC* plasmids; all but one, the dnaA protein, were familiar to us from their roles in other replication events. By manipulating the order of their addition and reaction conditions (temperature, time, salt, levels of magnesium and ATP, and so on), we were able to separate the overall reaction into several discrete stages. Simply stated, the origin of the *E. coli* chromosome, marked

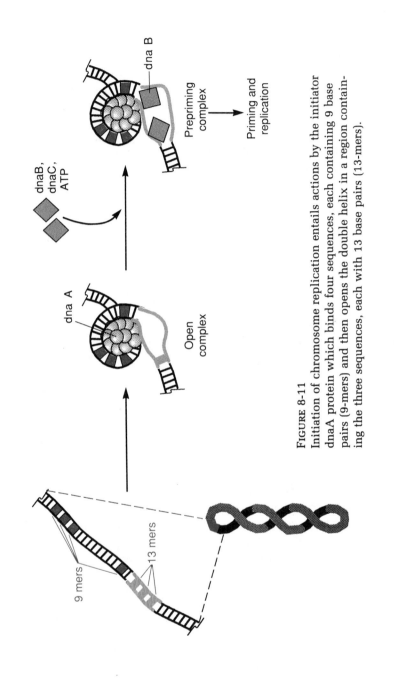

FIGURE 8-11

Initiation of chromosome replication entails actions by the initiator dnaA protein which binds four sequences, each containing 9 base pairs (9-mers) and then opens the double helix in a region containing the three sequences, each with 13 base pairs (13-mers).

and slightly opened by a special protein (dnaA), is then opened more in this susceptible region by another protein (dnaB), so that templates can be exposed for the synthesis of new DNA chains (Fig. 8-11). To begin with, some 20 dnaA protein molecules bind their sites in *oriC* and give it a special conformation. Then dnaB and another protein react with dnaA protein and unzipper the double helix in both directions from *oriC*. Opening up the duplex DNA permits primase to lay down primers from which DNA chains can be extended.

One feature of the reaction came as a big surprise. Immediately, when we started to dissect the crude enzyme mixture (Fraction II), we were shocked by the loss of the absolute requirement for the *oriC* sequence and the dependence on dnaA protein. Then we were relieved to find that the specificity observed with the cruder enzyme preparations could be recovered by adding back discarded protein fractions which had no replicative activity by themselves. We then set out to purify these "specificity" proteins.

After nearly a year's effort, we reached the end of our chase for the missing proteins. To our chagrin, they were two familiar enzymes, topoisomerase I and RNase H, already stocked in our freezer. The solace for having wasted so much effort was our familiarity with the properties of these enzymes. We could attribute their specificity functions to aborting initiations of replication on duplex DNA at locations other than *oriC*. We felt more secure in this interpretation when genetic results were later published on the behavior of mutant *E. coli* deficient in one or the other of these proteins: initiations of chromosome replication in these mutants took place elsewhere than at *oriC* and no longer required dnaA protein. Such mutants, incidentally, are not as "healthy" as normal cells and do not thrive under rich nutritive conditions that favor rapid growth.

We now have in hand the parts of a working switch that initiates replication at the origin of the *E. coli* chromosome. Surely, its operation in the cell must be under refined control, programmed to respond to many signals linking replication to nutrition, growth, and cell division. Understanding these decisive regulatory reactions is the challenge facing us at the next level of complexity. In approaching these problems we are, at this writing, discovering that we need to become more knowledgeable about the anatomy and sociology of dnaA protein, the leading actor in these opening scenes.

Cell Membranes and the Partition of Daughter Chromosomes

How plausible it seems that sorting of the completed pair of chromosomes between two daughter cells should be managed by their attachment to the cell membranes. A relatively simple mechanism can be imagined in which the membrane that divides the parental cell into the two daughter cells grows between two sites to which the chromosomes are attached, thereby ensuring their equipartition. Despite many experimental attempts during the twenty-three years since it was suggested by François Jacob, Sydney Brenner, and François Cuzin, this attractive scheme still lacks any firm evidence to support it.

What is a cell membrane? To the membrane that envelopes it and separates it from the environment, the cell owes its integrity and its very life. Even a cell such as the human red blood cell, which lacks DNA and protein-synthesis machinery, can still live for several months, metabolizing, respiring, and performing its vital oxygen-carrying functions, as long as its membrane is intact. But a tiny puncture in the membrane, causing leakage of cellular contents, may spell instant death. Among their many additional specialized functions, membranes likely have a role in sorting chromosomes between daughter cells, and we would like to understand how *E. coli* cells manage this partition.

The membranous envelope of *E. coli* is made up of four layers (Fig. 8-12):

(1) The *inner membrane* is comparable to the single (plasma) membrane that encircles an animal cell. It gives the cell its integrity by regulating the passage of molecules between the cellular interior and the environment. It is a flexible stockade of soap-like phospholipid molecules, anchoring and housing a variety of protein molecules, some of which, facing inward, might serve as attachment points for DNA.

(2) The *periplasm* (outside the inner membrane) is an antechamber, occupying nearly a third of the cell's volume. For lack of a digestive tract, many of the cell's digestive and nutritive functions are performed here.

(3) The *cell wall* (peptidoglycan layer) is a giant sack-like molecule that gives the cell its shape and rigidity and forms the

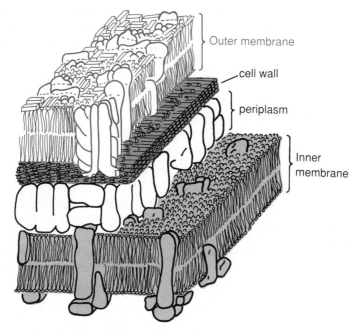

Outer membrane

cell wall

periplasm

Inner membrane

FIGURE 8-12
Multiple layers of the E. *coli* envelope include an outer membrane, a rigid cell wall, periplasm, and an inner membrane.

septum that divides one cell into two. Interference with the orderly growth of the cell wall is the basis of penicillin's antibiotic action. A resting bacterial cell is not disturbed by penicillin, but when the cell increases in volume prior to division, blockage of orderly wall growth and triggering of autolytic enzymes by the drug leads to membrane blisters with subsequent ruptures and cell death. Animal cells, because they lack this peptidoglycan envelope, are unaffected by these antibiotics.

(4) The *outer membrane* is the first barrier to transit of molecules from the outside. It also contains the sites to which viruses attach and possesses the chemical features (antigens) against which an animal host develops antibodies.

It would be simplistic to regard these cellular membranes as being arranged like the layers of an onion. Very likely there are organized patches in which the layers coalesce, affording anchoring points for stages in the replication cycle, and channels which viruses exploit to enter the cell. In order to expand my limited knowledge about membrane structure, growth, and functions, I have on several occasions studied the metabolism and properties of the phospholipids of which membranes are composed. This effort has been largely unrewarded. Yet I am once again working with membranes, hopeful that my increasing familiarity and concern with them will this time yield novel insights into the mechanics and control of DNA replication.

My first of several affairs with membranes, back in 1952, grew out of the mistaken notion that the synthesis of the relatively simple phospholipid molecule might be an instructive model for how the backbone of a complex nucleic acid chain is built. I thought synthesis of the phosphate bridge between the two "alcohol" moieties of a phospholipid (glycerol and choline) might teach us how the phosphate connection between the alcohol-like sugars of a nucleic acid backbone was made. Although this approach took me far off the mark, I learned a valuable lesson from it. In the future, I would rely less on my intuition about models and try to head toward an objective more directly. Yet, this misguided effort did lead to an exciting discovery.

Although the liver extract I was using failed to carry out the simple reaction I was after, I happened to find that it catalyzed the synthesis of the more complex phospholipids. At that time in 1952, as a candidate for a position at Massachusetts Institute of Technology, I proudly presented these findings in a seminar entitled "The Enzymatic Synthesis of a Phospholipid." After I finished, something very strange happened. Professor S. J. Thannhauser (1895–1962), the patriarch of phospholipid studies, rose to say: "I was worried when I saw the title of your talk that you had really discovered the enzymatic synthesis of phospholipids after we had all failed. Now I am relieved to learn that you have merely found the synthesis of phosphatidic acid. We all know that phosphatidic acid is a product of phospholipid breakdown. It is not detected among cellular phospholipids, and surely has no importance, even as an intermediate in their synthesis." The audience gasped, and my host, the late Gerhard Schmidt, a student of Thannhauser's in Germany, was covered with embarrassment. I countered by saying that phosphatidic acid is by definition a phospholipid, that I had also presented preliminary evidence for the synthesis of lecithin (a major phospholipid constituent of membranes), and, finally, that the apparent absence of phosphatidic acid in cells could be best explained by its extremely efficient use as an intermediate on the pathway to the other phospholipids.

I failed to get the MIT job, but I was right about the enzymatic synthesis of phospholipids. After I moved the next year from NIH to Washington University, I was too busy with the biosynthesis of nucleotides to continue the phospholipid work. Paul Berg, who had come as a postdoctoral fellow, was uninterested in phospholipids, but Eugene Kennedy, then at the University of Chicago, found my preliminary results helpful to him in his pursuit of phospholipid biosynthesis. In the following decades, as Kennedy's outstanding studies revealed a variety of fascinating turns in the byways of phospholipid biosynthesis, I had tinges of regret that I had missed out on the pleasures of some of these discoveries.

Finding DNA or a replication protein in a membrane fraction of a disrupted cell is commonly reported as proof of their functional connection; seeing things touch in a selected electron micrograph has been regarded as evidence of a specific interaction. Unfortunately, guilt by association is as common a miscarriage of justice in biochemistry as it is in the law. Such evidence of interaction must remain suspect because of the likelihood that two charged, sticky molecules which are functionally unconnected in the cell will attract each other adventitiously as a result of the massive turmoil required to make a crude cell juice. Among a number of possibilities of inner membrane involvement in DNA replication that arose in our work during the next fifteen years, none has stood up to rigorous tests of chemical and functional specificity. We wasted several years searching in a variety of membrane fractions for the enzyme system to replicate oriC plasmids; the vigorous enzyme system was free of membranes.

Innocence by lack of apparent association can be equally misleading. Functional connections may be lost due to the dilution, enzymatic attack, and chemical reactions that attend cellular disruption. Despite our many unsuccessful attempts, it still makes good chemical and biologic sense that several stages of chromosomal replication would be oriented and directed by attachment to the inner membrane. Recent developments in our work encourage me to think that these many years of tracking membrane involvement in replication have finally put us on to some strong scents.

We have found that dnaA protein, the crucial factor in starting a cycle of chromosome replication, has a remarkable affinity for certain phospholipids in *E. coli*'s inner membrane. This affinity is shown dramatically when dnaA protein, having bound an ATP molecule, hydrolyzes it to ADP, which is also bound firmly and renders the protein inactive in replication. Particular phospholipids are unique, among numerous agents and procedures tested, in their capacity to displace the ADP and rejuvenate the inert dnaA protein. This highly specific interaction between a key replication protein and membrane constituents known to undergo rapid changes in the cell cycle opens up exciting possibilities for the regulation of replication and cell division.

Plasmids offer an attractive approach to discovering how chromosomes are partitioned between the daughter cells. For these tiny DNA circles, one-thousandth the size and complexity of the chromosome, to survive within a bacterial population, it is essential that they be distributed evenly between the daughter cells at division. If the distribution of plasmids were left to chance, so that sometimes most or all would go into only one of the two daughter cells, eventually plasmids, which initially exist in only a few copies per cell, would dwindle in number and disappear after several cell generations. The survival of plasmids is achieved by various devices that ensure their equitable distribution in the host population. To probe the biochemistry of these mechanisms, assays are needed that monitor the transit of DNA molecules *between* cells. Compared with conventional assays that assess DNA transactions completed within the cell compartment, such biochemical determinations of *intercellular* traffic now present a novel and formidable challenge.

Cancer, Aging, and the Cell Cycle

"Doctor, how might your work on DNA help us with cancer?" This question is usually asked by one of those few who have remained alert at the end of my occasional "popular" lectures. My first reaction is usually disappointment that I have not captivated my audience with my descriptions of the overwhelming beauty of Nature's design for replicating its genetic material. But I can also appreciate the intense interest people have in how aberrations in replication might lead to cancer and aging.

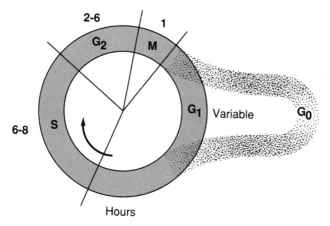

FIGURE 8-13
Cell cycle in animals and plants. A quiescent interval (G$_0$) is
followed by periods of growth (G$_1$), DNA replication (S), preparation for mitosis (G$_2$), and mitosis and cell division (M).

My recent answer to this question has been vague but not discouraging: Cancer is a broad term that includes a hundred different
diseases in which the normal controls over cellular growth have
been lost. For some cancers, we know the inciting agents and can
avoid them; in a few, the process can be arrested. I am impressed by
the increase in the past ten years in basic knowledge about the genes
and proteins responsible for the orderly growth and functions of
cells. Unless we squander our research resources, the pace of discoveries will be even greater in the next ten. I am hopeful that applications of this knowledge about the chemistry of genes and the way
they are replicated will in our lifetime significantly improve the
prevention, diagnosis, and treatment of many kinds of cancer.

As for aging, the problem is far more difficult than cancer, because
we simply don't know what aging is or how to measure it. In an adult
human, millions of cells die every second and must be replaced.
Most prominent among them are the cells that populate the blood,
make up the skin, and line the intestines. Cell division takes place in
four stages and consumes twenty-four hours on average (Fig. 8-13).
Ninety percent of the time is spent in a growth period (G$_1$) in preparation for a DNA replication stage (S) followed by another growth
interval (G$_2$); the mitosis phase (M), in which the duplicated DNA is

condensed into chromosomes and divided between the daughter cells, is brief.

The growth cycle is more easily examined in cells in culture than in tissues or tumors. When normal cells sense that their division and spreading on a surface is limited, they become arrested (G_0) before entering the (G_1) phase. Cancer cells ignore this barrier, keep dividing, and pile on top of one another; they appear to have lost a feedback control mechanism. When a population of normal cells, transferred periodically to a fresh culture medium, approaches fifty generations, the cells take longer to divide and eventually die. This "programmed cell death" is not observed with cancer cells; they seem to be immortal.

Our understanding of the biochemical basis of the control of cellular growth and its transformation into cancer is still woefully poor. We do not know how a signal received at the cell surface awakens an arrested cell to proceed on to DNA synthesis and cell division. Nor can we identify any of the mechanisms responsible for the uncontrolled proliferation of the cancer cell. Yet, there is reason to be hopeful that the current excitement about oncogenes will lead to significant insights into these processes.

Put simply, an *oncogene*, as part of a cancer virus, is the gene responsible for transforming cells to malignant growth in culture and in animals. More than fifty different oncogenes are now known. In each case, the oncogene or a very closely related gene (*proto-oncogene*) has been found in the chromosomes of normal cells. It may be the overabundance of this cellular gene, its misuse, or its mutation that perverts the normal pattern of cellular growth. Some of the oncogenes code for proteins that are membrane-bound and have the capacity to transfer phosphate from ATP to the amino acid tyrosine in specific cell proteins. Despite the many examples and studies of this property, its significance remains utterly mysterious.

The huge activity in oncogene research, with papers numbering in the thousands, needs to be supplemented with knowledge of the mechanisms and control of DNA replication and the expression of its genetic information in animal cells. When this gap in basic biochemistry is narrowed and these facts are coupled with the enormous power of genetics and genetic engineering, significant and gratifying progress will be made in understanding cancer and the aging process.

Chapter 9

Gene Hunters and the Golden Age

In this golden age of genetic chemistry and recombinant DNA, many wonder where it all began and who can lay the best claim to it. Assuming that molecular biologists are dominant among the gene hunters, where did they come from? Physics, chemistry, biology, or biochemistry?

As a biochemist (but one who has frequently been identified as a molecular biologist), it seems to me that the physicists, geneticists, and their admirers who chronicle the origins and exploits of molecular biology have ignored the major contributions of biochemistry. Their collective influence has had far-reaching implications. Research and teaching staffs of many departments, including biochemistry, have been depopulated of their biochemists and now lack the philosophical outlook and technologies essential for progress toward the goals of molecular biology.

Biochemistry, with roots in organic chemistry, medicine, and agriculture, goes back more than a hundred years and embraces the objectives and practices of molecular biology, a discipline less than half its age. The description of biochemistry which I like best was given by F. G. Hopkins in 1931. I recited it each time I gave the opening lecture of the introductory course at Stanford:

> The task of the biochemist wishing to get to the heart of his problem is exceptional in that he must study systems in which the organization of chemical events counts for more, and is carried far beyond such simpler coordinations as may be found in non-living systems . . . Current philosophy is busy in emphasizing the truism that the properties of the whole do not merely summarize but emerge from the properties of its parts, and some exponents hold *a priori* that biochemical data can throw no real light on the nature of an organism which, in its very essence, is a unit. The biologist has long studied living organisms as wholes and will continue to do so with ever-increasing interest. But these studies can tell us nothing of the nature of the "physical basis of life," which no form of philosophy can ignore. It is for chemistry and physics to replace the vague concept "protoplasm"—a pure abstraction—by something more real and descriptive. I know of nothing which has shown that current efforts to this end do not deal with realities. It is only necessary for the biochemist to remember that his data gain their full significance only when he can relate them with the activities of the organism as a whole. He should be bold in experiment but cautious in his claims. His may not be the last word in the description of life, but without his help the last word will never be said.

Whereas molecular biology has dwelt largely on the triad of DNA, RNA, and protein, and the transfer of genetic information between them, biochemistry is concerned with all the molecules of the cell and organism. Excluded from the province of molecular biology have been most of the structures and functions essential for growth and maintenance: carbohydrates, coenzymes, lipids, and membranes; the processes of energetics and metabolism; motor and special sensory functions.

Traditionally, the biochemist pursues a *function* (such as fermentation of sugar to alcohol, photosynthesis, vision, replication of DNA) to discover the structures responsible for it. He feels impelled to break open the cell and devise assays in order to purify the molecules that reproduce the cellular function. The biochemist focuses on chemistry to connect genetics and physiology with the physical features of a system. By contrast, the molecular biologist commonly pursues a *structure* (DNA or an available protein) to find its functions. He modifies the structure, introduces it into an intact cell, and from the cellular responses he tries to infer the functions of the structure. Often trained in physics or genetics, he may feel uneasy with biochemistry and generally avoid it.

In their discovery of the DNA double helix in 1953, Francis Crick and James Watson hit the biological jackpot of the century. The dazzling success of the double-helical structure lay not so much in accounting for the multitude of physical and chemical properties of DNA as in providing compelling mechanisms for the replication of DNA, its mutability, and, later, the expression of its genetic information. This discovery of the double helix, built upon decades of biochemical research, catapulted molecular biology to a status that eclipsed the traditional practices of biochemistry.

Biochemistry and the Origins of Molecular Biology

In trying to redress the neglect of the biochemical roots of molecular biology, Seymour S. Cohen, a pioneer in the application of biochemistry to the study of viruses, goes back to 1850, to the physiologist Claude Bernard. Around that time in France, Bernard showed the digestion of fat to be a distinctive chemical hydrolysis located in a specific region of the intestine. Based on analyses of several digestive and metabolic events, he established causal connections between a chemical reaction and a vital process. In so doing, he was effecting a transition from general physiology, the study of mechanisms of biological functions, to a sharper focus on the isolation and characterization of substances and reactions in the body, a discipline which came to be known as physiological chemistry, and still later as biochemistry.

An early achievement in physiological chemistry was the discovery of nucleic acids by Friedrich Miescher in 1869. He had been admonished by his uncle, Wilhelm His, the famous anatomist, not to study anatomy but to explore the chemical composition of tissues "because the ultimate problems of tissue development would be solved on the basis of chemistry." To examine the composition of cell nuclei, Miescher made extracts of the pus on surgical bandages, an abundant source of nucleated cells. From these extracts he obtained a strange substance rich in phosphorus. Soon after, he isolated a similar substance from salmon sperm. He called it nuclein, later renamed nucleic acid. Several decades of chemical analysis of nucleic acids from plants and animals identified their nucleotide components and distinguished RNA from DNA. (In 1938, when I first studied biochemistry, RNA was regarded as the nucleic acid of plants and DNA as that of animals; moreover, a nucleic acid was believed to be a chain of tetranucleotides!)

In the two decades before Watson and Crick developed their model for DNA, a series of biochemical discoveries made several things clear: RNA and DNA are in all cells, plant, animal, and bacterial; nucleic acids are very long chains of nucleotides; the heritable organism which causes mosaic disease in plants—a virus—is simply a molecule of RNA wrapped in a protein sheath; and DNA is the genetic substance of all creatures—microbial, plant, and animal. Identifying the DNA molecule as the chemical basis of genes and heredity in 1944 culminated Oswald Avery's life-long obsession with pneumonia and the agent that causes it, the pneumococcus. In concentrating his attention on these bacteria, he eventually grappled with the phenomenal capacity of strains to "transform" one another. (In 1928 Fred Griffith, 1877–1941, had discovered that, in a mouse infected with two strains of pneumococcal bacteria, one bacterial strain can acquire the hereditary characteristics of the other.) Avery tracked this transforming capacity to extracts of the bacteria and finally showed that DNA is the molecule that endows each strain with its distinctive features.

In analyzing the composition of DNAs from different species, Erwin Chargaff found in all DNAs that the percentage of A equaled that of T and the percentage of G equaled that of C; however, the ratio of A plus T to G plus C varied widely and could be regarded as the signature of each species. When to all these biochemical facts was added the evidence from Linus Pauling's work that protein chains are helically shaped, virtually all the clues were there. In a close four-team race, Watson and Crick were the first to fit the available x-ray analyses of DNA with a double-helix model. The opposite orientation of the two chains in the double helix could not be deduced from these data, however, and was established only eight years later from our enzymatic studies of DNA synthesis.

Another strong but recent root of molecular biology developed from the intensive study of the biology and biochemistry of the E. coli phages. Starting around 1945, under the leadership of Max Delbrück and enveloping other expatriate physicists, the "phage group" at Cold Spring Harbor Laboratory on Long Island focused intensively on how the T2 virus managed in twenty minutes to produce two hundred copies of itself. "Surely," mused Delbrück, "a group of smart people should, with concerted effort, clarify in a few years a biologic process that takes only a few minutes." Results of simple manipulations of the genetics and physiology of populations of phages and bacteria could be observed in Petri dishes and expected to give decisive answers. There would be no messing around with cell extracts and related biochemical mumbo-jumbo.

When the success of molecular biology was celebrated in 1966 in a *Festschrift* for Max Delbrück, phages were given full credit for the emergence of this new discipline. Under the title *Phage and the Origins of Molecular Biology*, editors John Cairns, Gunther Stent, and James Watson collected a series of impressive papers that illustrated the major contribution phage studies had made to molecular biology. Unfortunately, omission of the earlier and sustained contributions of biochemistry and physical studies of DNA and proteins left the distorted impression that phages were the sole source of molecular biology. Included in the volume was the classic biochemical demonstration by Alfred Hershey (a maverick cardinal in the Delbrück church) that the DNA injected into the cell by the syringe-like phage is responsible for the hereditary information, and not the protein left outside. Yet, the important biochemical phage studies that provided the foundations for the Hershey experiment and the highly significant ones that followed were ignored.

Robert C. Olby, the eminent historian of molecular biology, agrees with biochemists that biochemistry has been unjustly treated by Stent and Watson in their popular histories of molecular biology. But he also implies that the unreceptive thinking of biochemists at various times may have been in part responsible. Though exhuming opinions from here and there can be a pointless exercise, I do recall that during my stay in St. Louis in 1947, Gerty Cori, the quintessential biochemist, gave me the Avery paper and said: "You must read this. It is very important." Yet five years later, the celebrated phage genetics group at the California Institute of Technology, including Delbrück, still ignored the Avery discovery of DNA as the chemical basis of heredity.

It seems to me that the essence of molecular biology as it is currently practiced is biochemistry. Yet most of the molecular biologists who practice this specialized form of chemistry do not regard it so.

They identify and isolate a tiny gene from relatively huge chromosomes, often only one part in millions or billions, and then they amplify that part by even larger magnitudes using biochemical and microbial cloning procedures. These techniques, known collectively as genetic engineering, enable them to map chromosomes, redesign their genetic arrangement, minutely analyze their composition, isolate their components, and produce these components in bacterial factories on a massive industrial scale. Not even the boldest among us dreamed of this kind of chemistry ten years ago.

Yet with all its success, molecular biology has not solved some of the profound questions of cellular function and development. What governs the rearrangements of genes that allow them to produce antibodies? What determines whether a primitive cell will develop into a brain cell or a bone cell? What process underlies the growth and senescence of cells? Current approaches falter when they ignore the chemistry of the products of the DNA blueprint—the enzymes and proteins that represent the machinery and framework of the cell.

The tides of fashion in science erode one beach to create another. In the rush and excitement of the new mastery over DNA, training in enzymology and its practice have been neglected. Most of our students are introduced to enzymes as commercial reagents, and they find enzymes as faceless as buffers and salts. Biochemists deserve to share in the glory of the golden age of discoveries of genetic chemistry and molecular immunology often attributed to others who are indifferent to biochemistry.

Despite their limitations, Nobel Prizes in science are a measure of the novelty and significance of discoveries. Lars Ernster, a member of the chemistry selection committee for many years, analyzed the awards in biology (medicine or physiology) and chemistry since their inception at the start of this century. The result is remarkable. Forty percent of the awards in both biology and chemistry have been in the *hybrid* discipline of biochemistry.

In recent years, the popularity and power of genetic engineering, dominated by molecular biology, have eroded the ranks and important roles of biochemistry. Without attention to biochemistry, the basic issues of cell growth and development, of degenerative disease and aging, will not be resolved. Molecular biology has successfully broken into the bank of cellular chemistry, but biochemical tools and training are needed to unlock the major vaults.

Origins of Recombinant DNA

Even if the genealogy of molecular biology were settled, no one could have predicted from its antecedents the stupendous power of this offspring. Gunther Stent, a notable philosopher and historian of science, pronounced the death of the golden age of gene hunting when it had barely been born. Cascades of technological advances produced a torrent that made engineering of genes, chromosomes, and species an everyday laboratory exercise. The origins of recombinant DNA techniques are still so fresh and close to home that they deserve a brief review before the events and those who made them have faded from view. None of us working for so many years on the enzymes that make and break DNA ever imagined they would become the indispensable tools of the most sweeping technical advance in the history of biology. Nor did any of us anticipate that this tidal wave of genetic engineering would come so quickly and envelop everything about us.

The power of genetic engineering is its utter simplicity and wide applicability. Perceived by business as a financial blockbuster and indispensable to progress in medicine, agriculture, and chemical industry—not unlike the rage for plastics and computers in preceding decades—hundreds of enterprises are now competing to sharpen the tools and devise novel ways to use them. A sizeable publishing industry has sprung up simply to report and advertise these developments.

Significant roots of genetic engineering grew in the Biochemistry Department at Stanford because we had discovered or were applying the reagents basic to the manipulation of DNA: polymerase to synthesize long chains of DNA and fill in gaps, ligase to join contiguous ends of chains, exonuclease III to remove obstructive phosphate groups at chain ends, phage λ exonuclease to chew back one end of a DNA chain, and terminal transferase, an enzyme (from the thymus gland) to add nucleotides, willy-nilly, to the other end of a DNA chain. These five enzymes were among the reagents that stimulated and nourished the two Stanford experiments that introduced recombinant DNA technology and led to the engineering of genes and chromosomes.

One of these experiments was the graduate thesis of Peter Lobban, whose faculty advisor was Dale Kaiser. Lobban had worked with phage λ as an undergraduate at MIT but had also taken a variety of engineering courses. After three years at Stanford, his projects on λ had not yet gelled, and he spent the summer of 1969 fulfilling one of the rather minimal departmental requirements, a written proposal for research in an area outside his immediate interests.

Lobban had been impressed by the ability of infecting phage DNA to recombine with the host chromosome and to extract specific genes when it emerged from it. With his strong appetite for engineering, he wondered if this kind of recombination could be managed outside the cell. He laid out a detailed plan for how the ends of two different DNA molecules (such as a human insulin gene and phage λ DNA) might be tailored to make them right for joining. The recombinant DNA could be smuggled into *E. coli* using a Kaiser–Hogness technique in which such entry is helped by a resident phage.

The proposal was as follows (Fig. 9-1):

(1) Add a tail of A nucleotides (using terminal transferase) to each of the 3 ends of the chains of one piece of double-stranded DNA and similar tails of T nucleotides to the 3 ends of the chains of another batch of DNA. (Suspecting that this enzyme has a strong preference for single-stranded DNA, the 5 ends of the chains would first be shaved back a bit with the phage λ exonuclease that can do this job.)

(2) Mix the two populations of tailed molecules; hybrids will form through annealing of their A and T tails. (As Lobban later recalled, the idea of using molecules tailed by terminal transferase to generate recombinant DNA came to him from a seminar given by a fellow graduate student, Tom Broker. At the time, Broker was working under Bob Lehman's direction on recombinant molecules produced in cells infected with T4 phage.)

(3) Separate the circular hybrid (nearly twice the length of the initial DNA) from the tailed linear pieces and other forms of DNA by its more rapid and distinctive sedimentation in a centrifuge.

(4) Fill gaps in the DNA and repair raw ends with polymerase; join the contiguous chain ends with ligase to form a complete duplex circle.

(5) Use available electron microscopic and chemical methods to verify each step in the end-to-end joining of two different DNA molecules.

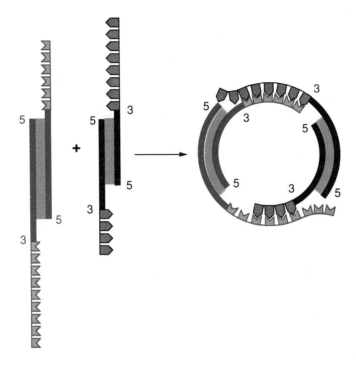

FIGURE 9-1
The first technique for making recombinant DNA. A tail of T nucleotides
has been added to the 3 end of each of the strands of a piece of double-
stranded DNA from one source, and a tail of A nucleotides has been similarly added to DNA from another source. (The tails have been simplified
to show only the bases; the size of the tail has been enormously exaggerated for this illustration.) Molecules of the two populations are mixed and
become joined by the matching of T tails to A tails. Gaps can be filled by
synthesis with DNA polymerase; contiguous chain ends are then joined
by ligase.

    Bob Lehman, Buzz Baldwin, and Dale Kaiser—the faculty group
that heard Lobban's proposal—liked the idea and thought the experimental plan so feasible that they encouraged him to try it. Kaiser,
knowing that Paul Berg's group down the hall had been thinking
along similar lines, urged Peter to check with them before starting.
    A year or so earlier, Paul Berg became fascinated with the simian
virus (SV40) that infects and produces tumors in monkeys. Berg felt
that SV40, like the tiny phages of *E. coli*, had the virtues of simplic-

FIGURE 9-2
Catenanes, interlocked DNA circles.

ity and held the promise that the study of the viral life cycle would illuminate the mechanisms used by the animal cell to replicate its own DNA and to express its genetic information. Also, the susceptibility of cells to infection with free SV40 DNA makes it possible to observe the consequences of experimental dissections and rearrangements of the DNA made outside a cell.

Having heard a seminar by James C. Wang about interlocked DNA rings (catenanes; Fig. 9-2), Paul wondered whether the SV40 DNA circle might be used in this way to ferry foreign genes into an animal cell. For lack of methods to catenate DNA molecules, David Jackson and Bob Symons, two postdoctoral fellows working with Paul, were considering other ways of linking foreign DNA to SV40, among them the A-T tailing method. They had an additional problem. Because SV40 was a circle, it had first to be opened to attach the foreign DNA and then be recircularized.

At that time, the goal of Lobban and Kaiser to make *E. coli* a factory for the amplification and expression of mammalian genes by using a phage as a vector seemed distinct from that of the Berg group, which was to introduce foreign genes into mammalian cells using SV40 virus as a vector. The two groups met and agreed to pursue their separate directions and to keep in touch. Lobban's methodical studies of the transferase and nucleases and his meticulous use of them were crucial to the success of the recombinant DNA procedure for both groups.

Cleavage of the SV40 circle at a unique place was accomplished with a nuclease called EcoR1, which became available just at the right time. This enzyme is one of a class of enzymes—restriction

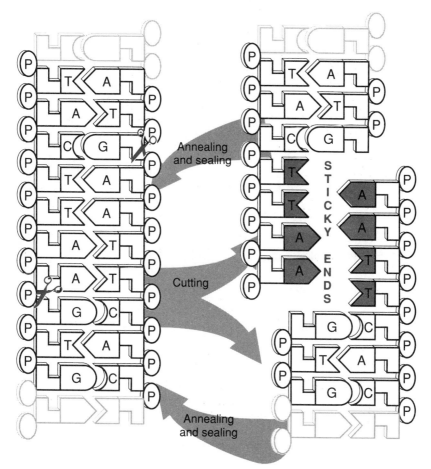

FIGURE 9-3

A particular restriction nuclease cuts DNA molecules only at a particular sequence of nucleotides to create protruding ends. By annealing these "sticky" ends, DNA molcules form lap joints, which can be sealed by ligase.

nucleases—of which hundreds have now been discovered, that recognize a specific sequence of four or more nucleotides and cut the DNA precisely within that sequence but nowhere else (Fig. 9-3). The discovery in the late 1960s of the awesome ability of these enzymes to carve massive chromosomes into manipulable pieces made them indispensable tools of genetic engineering. It also made Nobel Laureates of Werner Arber, Hamilton Smith, and Daniel Nathans in

1978. John Morrow, a graduate student of Berg's, uncovered the fortunate fact that the sequence cut by EcoR1 occurs only once among the 5,224 base pairs of SV40, and would soon figure again in an important event in recombinant DNA history, as would EcoR1.

In May 1972 Lobban submitted his 137-page PhD thesis describing the joining of two linear DNA molecules to form a circle essentially as outlined in his 1969 proposal; he used a λ-related phage, P22, because its varied DNA ends provided a more stringent test for joining different DNA molecules. By July 1972 Jackson, Symons, and Berg were preparing a manuscript for publication describing their successful procedure for inserting genes from phage λ and *E. coli* into an SV40 circle. After cleavage of the SV40 circle by EcoR1, they had attached A tails to its ends and T tails to the DNA bearing the phage and bacterial genes. They mixed the A-tailed with the T-tailed molecules to let them anneal and then processed them into a double-length circle much as Lobban and Kaiser had been able to do in their end-to-end joining of linear P22 DNA (Fig 9-1).

It seemed reasonable that brief reports of the success of both groups should appear together in an early issue of the *Proceedings of the National Academy of Sciences*. Kaiser preferred, however, that some loose ends in Lobban's thesis work be completed. Thus, the extensive account of the P22 work submitted to the *Journal of Molecular Biology* in February 1973 did not appear until nearly a year after the Berg group's historic paper. In retrospect, this decision to wait may have led to events which were responsible for Lobban's being diverted from a career in science.

Among the pioneers of recombinant DNA, Peter Lobban is not only unsung but is no longer even in the chorus of basic science. In 1972 Lobban went as a postdoctoral fellow to the University of Toronto to learn animal cell genetics from Lou Siminovitch, with the objective of introducing animal genes into bacterial cells. Two years later, when he looked for an academic appointment, he failed to get an offer from any of the fifteen universities to which he had been invited for an interview. Why? He chose for a seminar subject his incomplete postdoctoral work rather than his earlier achievements on recombinant DNA at Stanford; and even if he had chosen recombinant DNA, the scientific community might not have then appreciated the significance of this new technology. Moreover, rumors of a moratorium on genetic engineering research were rife, at a time when there was a glut of qualified applicants. Also, Peter is diffident and shy and had always felt comfortable with engineering as an alternative career. He returned to Stanford to take a master's degree in electrical engineering and now manages the computer programming at a medium-sized medical instrument company in Silicon Valley.

Genetic Engineering and the Marketing of DNA

In the drama of genetic engineering, the EcoR1 restriction nuclease was an early addition to the cast of star performers. Herbert Boyer recognized the enzyme as the product of a gene in an *E. coli* plasmid called R1, hence the name EcoR1. After John Morrow found that EcoR1 cleaved a single site in SV40 DNA, Janet Mertz, another Berg graduate student next door, teamed up with Ron Davis, a newly arrived faculty member, to show in 1972 that the breaks it made in the two DNA chains were staggered (Fig. 9-3). As a consequence of the staggered cleavage by EcoR1 nuclease, the two products have matching ends. Such complementary or cohesive ("sticky") ends make it easy for them to be reannealed and aligned as a lap joint so that their chains can be joined by ligase. They showed further that all DNAs cleaved by EcoR1 have the same ends and can thus be recombined by the sequential action of EcoR1 and ligase to generate the recombinant DNA desired.

Stanley N. Cohen, a physician in the Department of Medicine at Stanford who had been concerned about bacterial resistance to antibiotic treatments, had been doing research on plasmids, the prime source of the resistance problem. Having had biochemical training, he had close working relations with members of the Biochemistry Department and was attentive to the progress being made with recombinant DNA, particularly the latest development of using EcoR1 nuclease to make sticky ends and ligase to seal chain breaks. Boyer was also very knowledgeable about restriction enzymes encoded by plasmids, EcoR1 being his favorite. Cohen and Boyer both saw and exploited the great opportunity of using EcoR1 and ligase to create a chimeric plasmid, a hybrid endowed by one parent plasmid with the gene for resistance to one antibiotic (such as penicillin) and endowed by the other parent plasmid with the gene for resistance to another antibiotic (such as tetracycline).

They were also aware that a major obstacle to getting naked DNA into *E. coli* cells had just been breached. For over twenty years, many hundreds of investigators working in bacterial genetics had been frustrated by the reluctance of *E. coli* to imbibe DNA. Some other bacteria (*Pneumococcus, Hemophilus,* and *Bacillus*) did so readily and had by default become the favored objects for studying transformation, the transfer of genetic traits via DNA from one bacterial strain to another. Ironically, *E. coli,* in which Joshua Lederberg had first discovered bacterial mating and in which 99 percent of the subsequent genetic and biochemical effort had been invested, remained impregnable to the entry of free DNA. That is, until Mort Mandel discovered the use of calcium in 1971, a contribution that has received little acclaim but which was crucial to the history and development of genetic engineering.

While a member of the physics faculty at Stanford, Mandel had become intrigued by DNA in a seminar I had given in his department and decided to switch to biology. With Lederberg's sponsorship, he was doing spectroscopic work in the Genetics Department. Being next door to the Biochemistry Department, he became aware of the Kaiser–Hogness technique for introducing phage DNA into *E. coli,* learned how to handle phages during a fellowship year in Stockholm, and then joined the faculty of the University of Hawaii. He knew that calcium improved the infectivity of some phages and soon discovered that calcium can affect the membrane barriers of *E. coli* cells to promote a limited but adequate entry of DNA molecules. This trick is now an indispensable maneuver in virtually all genetic engineering operations.

Cohen and Boyer, with the assistance of colleagues, were immediately successful in using EcoR1 cleavage and subsequent ligase sealing to generate a recombinant DNA molecule from two plasmids—one, from *E. coli,* bearing the gene for resistance to tetracycline; the other, from *Staphylococcus,* carrying the gene for penicillin resistance. With the Mandel calcium trick to open the gates, they introduced the novel plasmid into an *E. coli* culture and made the cells permanently resistant to both antibiotics. Soon after, Morrow teamed up with the Cohen and Boyer groups to insert a gene from the toad *Xenopus laevis* into a plasmid. After introduction of the plasmid into *E. coli,* the animal gene was perpetuated in the bacterium as if it belonged there. The road had been opened wide for the introduction and propagation in *E. coli* of any other gene—bacterial, viral, plant, or mammalian—into what was apparently a most welcoming host. With their perception, knowledge, and drive, Cohen and Boyer were poised to apply this technology to industrial objectives.

Despite Cohen's reluctance, Stanford University and the University of California (San Francisco) patented his and Boyer's invention and use of recombinant DNA. Today, nearly a hundred companies each pay $10,000 a year for license to use the technique; should significant sales of recombinant DNA products result, these companies are obligated to pay additional amounts in royalties. Of Stanford's share of this income, one-third is administered by Cohen, one-third is administered by the Department of Medicine and the Department of Genetics (of which he was until recently the chairman), and one-third is retained by the university. The Department of Biochemistry and the many people in it who did much to make recombinant DNA possible get credit for their contributions but, because of squeamishness about patents and commercialization, get none of the cash.

As a founder of Genentech, the first and most successful of the genetic engineering companies, Herb Boyer's entrepreneurial skills and vision stand in stark contrast with my own. For many years my intimate friend Alex Zaffaroni, president of Syntex Research and the founder of ALZA Corporation, kept asking me about the feasibility of applying our emerging knowledge of genetic chemistry to the pharmaceutical industry. I kept telling him it was grossly premature—only to be proven wrong by Genentech's success, which was soon apparent to Wall Street and the rest of us. I did not anticipate that *E. coli* cells would be so permissive and could be made into factories for the production of massive amounts of precious hormones, vaccines, and interferons. I never doubted one day that the universality of DNA replication and expression would be exploited, but my timing was way off.

Understanding Life as Chemistry

The overworked word "revolutionary" does not exaggerate the extraordinary impact of genetic chemistry and engineering on science, medicine, and industry. Equally revolutionary, though unnoticed, is the more fundamental coalescence of the sciences basic to medicine.

This confluence lacks a name; it has no flag nor obvious commercial applications. Yet its effects will prove to be as profound as those of genetic engineering and the earlier famous safaris in search of microbes, vitamins, and enzymes.

In Pittsburgh in 1981, speaking at the dedication of a Convention Center, I chose the title: "Confluence of the Medical Sciences." At that site, where the Allegheny and Monongahela Rivers meet to form the Ohio, I thought of how human events might resemble the course of rivers. "A river meanders, gathers small streams, widens, deepens and may even split into two rivers that go their separate ways. On occasion, two rivers merge to make a more significant statement, a mightier river." It seemed to me that genetic engineering, created by the confluence of biochemistry, genetics, and microbiology, was that mightier river and would profoundly alter the course of science and medicine.

When I was a medical student in the late 1930s there was no conception of the chemical basis of heredity. Genetics was expressed in abstract terms. It seemed inconceivable to me then that the nature of inheritance would be explained in simple chemical language in my lifetime. Relativity and quantum mechanics had revolutionized physics and chemistry early in this century. I did not imagine that my generation would witness another such revolution. That expectation has proven largely true for physics and chemistry but not for biology. Today we know the precise formula and organization of DNA—the genes and chromosomes. We understand how DNA directs the functions and embodies the genetic information of all living things.

Another major confluence was the merging of immunology with genetics and biochemistry. It led to a most remarkable breakthrough—the monoclonal antibody technique, which creates biological factories that produce antibodies of uniform composition in huge quantities. Because of their specificity, antibodies are of immense practical value: they confer immunity against viruses, microbes, and cancer cells; they are useful for exquisitely sensitive detection in diagnosis and in tissue typing for transplantation; and they provide powerful tools for the isolation of enzymes, hormones, and other molecules of vital importance in medicine and industry.

In the century preceding World War II, several new scientific disciplines basic to medicine were developed on the foundations of physics, chemistry, and biology. My preclinical medical school curriculum just before the war consisted of seven discrete subjects: gross anatomy, microscopic anatomy, biochemistry, physiology, pharmacology, bacteriology, and pathology. Each of these subjects was taught by a separate department, as sharply bounded as the undergraduate departments of mathematics, classics, history, and geology were and still are.

Today the boundaries between medical and biological disciplines have vanished. The basic medical sciences have converged into a single discipline that makes research and teaching in all the departments interdependent and virtually indistinguishable. We see anatomy and its embryological development in the assembly of molecules into larger and larger entities that finally give rise to cells, tissues, and organisms. The physical and chemical properties of molecules and cells are intertwined to give a dynamic picture of the structure and function of the animal from brain to toe.

In an anatomy department, biologists, chemists, and physicists can present the human body to medical students as an uninterrupted ascent from atoms to man: from the tens of atoms that make a small molecule, to the thousands of molecules that make a polymer (such as a protein or a nucleic acid), to the millions of such polymers that make a cell, to the billions of cells that make a tissue, and the trillions of specialized cells that create a body. In a wider, panoramic view, the human body and its behavior becomes a tiny decoration in the tapestry of life interwoven with the incredible variety of plasmids, viruses, bacteria, plants, and animals in a 4-billion-year evolutionary development.

The coalescence of the biological and medical sciences, as illustrated by these several examples, is based largely on their expression in a common language, chemistry. Understanding all of life, human behavior included, as chemistry links the biological sciences with all the physical sciences—the atmospheric, earth, and engineering sciences. Chemical language, rich and fascinating, paints images of great aesthetic beauty with beguilingly deeper mysteries.

Yet the importance of chemistry as the foundation of all medical science is usually submerged and obscured by attention to specific and urgent problems. This simple truth may escape both physicians and scientists.

Physicians are inclined to action. There is the story, often retold, of the surgeon who, while jogging around a lake, spotted a man drowning. He pulled off his clothes, dove in, dragged the victim ashore, and resuscitated him. He resumed his jogging, only to see another man drowning. After he dragged the second one out and got him breathing again, he wearily resumed his jogging. Soon he saw several more drowning. He also saw a professor of biochemistry nearby, absorbed in thought. He called to the biochemist to go after one while he went after another. When the biochemist was slow to respond, he asked him why he wasn't doing something. The biochemist said: "I am doing something. I'm desperately trying to figure out who's throwing all these people in the lake."

This parable is not intended to convey a disregard of fundamental issues among physicians nor a callousness among scientists. Rather, it portrays the reality that, in the war on disease, some must contribute their special skills to the distressed individual while others must try to gain the broad knowledge base necessary to outwit both present and future enemies.

Among scientists, both physicists and biologists have had a low regard for chemistry and little patience for it. Those physicists who, during the post-World War II period, turned their attention to biology generally focused on genetics and hoped they could avoid biochemistry. Biologists know that enzymes determine the shape, function, and fate of cells, but they shudder at the multiplicity and chemical complexity of enzymes and try to ignore biochemistry. (In fact, the discipline called "biophysics" sometimes has been a refuge for biologists and physicists who feel uneasy with biochemistry.)

As for the general public, distinctions among the natural sciences have little meaning. Many people cannot distinguish between an atom and a molecule, a virus and a bacterium, or a gene and a chromosome. No wonder, then, that journalists and lawmakers keep asking us whether a molecule, a virus, or a cell is living. They become impatient when we fail to give them a straight and simple answer as to where and when life begins and ends.

"Better things for better living—through chemistry" was the slogan of a DuPont advertising campaign for many years. The purpose of the slogan's message was to inform the public of the value of plastics, herbicides, and industrial chemicals for our individual and collective well-being. The campaign was successful for a time in promoting goodwill for chemistry and the DuPont company. Then, "through chemistry" was dropped when it became widely known that these chemicals—like all things, natural or man-made—can be toxic, too. In a foodmarket not too long ago, I overheard a little girl saying: "Mummy, you shouldn't buy that. It has chemicals." Neither the child's mother nor the store clerk found her remark strange or disturbing. No advertising campaign, nor our educational system, nor the media, including the excellent *Nova* television programs, have taught the public that life is a chemical process. These efforts have not made it clear that the human organism, its form and behavior, are determined by discrete chemical reactions, as are its origin, its interactions with the environment, and its fate.

Will the impressive advances that have been made in understanding the chemistry of metabolism underlying such diseases as diabetes, ulcers, and heart disease lead the public and physicians to believe that we can also come close to understanding the mind and human behavior? The first and most formidable hurdle is acceptance without reservation that the form and function of the brain and nervous system are simply chemistry. I am astonished that otherwise intelligent and informed people, including physicians, are reluctant to believe that mind *is* matter and *only* matter. Perhaps the repeated failures of science to analyze social, economic, and political systems has perpetuated the notion that individual human behavior cannot be explained by physical laws.

Brain chemistry may be novel and very complex, but it is expressed in the familiar elements of carbon, nitrogen, oxygen, hydrogen, phosphorus, and sulfur that constitute the rest of the body. Brain cells have the same DNA that all cells do; the basic enzyme patterns are those found elsewhere in the body. Hormones once thought to be unique to the brain are now known to be produced in the gut, ovary, other tissues, and even in plants and protozoa.

My plea is for research on brain chemistry, in animals and humans, normal and sick. With the application of simple, well-known biochemical techniques, specific brain functions can be mapped and assayed. Further advances will come rapidly when additional chemical techniques are developed specifically to explore the nervous system. When the neurobiologists (the "head hunters") in the next decades explore the special sensory regions, motor areas, emotion centers, and vast uncharted tracts, we will see astonishing revelations about memory, learning, personality, sleep, and the control of mental illness.

To sum up, much of life can be understood in rational terms if expressed in the language of chemistry. It is an international language, a language for all time, a language that explains where we came from, what we are, and where the physical world will allow us to go. Unfortunately, the full use of this language to understand life processes is hindered by a gulf that separates chemistry from biology. I addressed this subject in 1987 in a talk to the American Association for the Advancement of Science. The title was "The Two Cultures: Chemistry and Biology." The gulf between these two cultures is not nearly as wide as the one between the humanities and sciences to which C. P. Snow paid particular attention. Yet, the fields of chemistry and biology remain two distinctive cultures, and the rift between them is serious, generally unappreciated, and counterproductive. We have the paradox of the two cultures growing farther apart even as they discover more common ground in genetic chemistry and recombinant DNA.

Chemists need to be aware that three billion years of cellular evolution have perfected molecules and molecular societies of awesome chemical sophistication. The chemist interested in catalysis, stereochemical specificity, polymer structure, metallo-organic reactions, surface effects, and a hundred other facets of chemistry will be challenged and instructed by the myriad of forms in biology. At a social level, there is an urgent need to inform our citizenry of the importance of chemistry. Understanding the chemistry of heredity and nutrition, of health and disease, is an intimate and compelling means to achieve this.

Biologists need to be aware that life processes, their evolution and variety, can be and ultimately must be described in molecular terms. They must resort to chemical techniques to refine and broaden the scope of their explorations of how plant, animal, and microbial forms compete and cooperate on earth. Only in chemistry will they find the linkage between cosmological events, the earthly origins of life, and the possible fates that await it.

In the long view, these cultural differences between chemistry and biology are dwarfed by an overall devotion to the larger culture of science. It is the discipline of science that enables ordinary people, whether chemists or biologists, to do the ordinary things which, when assembled, reveal the *extraordinary* intricacies and awesome beauties of nature. Science permits them not only to contribute to grand enterprises but offers them a changing and endless frontier for exploration. I like the remark of Einstein, who has proven as quotable in philosophy as in physics. He once said: "The most beautiful experience we can have is the mysterious. It is the fundamental emotion that stands at the cradle of true art and true science. Whoever does not know it and can no longer wonder, no longer marvel, is as good as dead."

Biotechnology: Biology or Technology?

Genetic engineering is a newborn baby that fascinates Wall Street and preoccupies academia. Who begot this child? Who named it? What can we do to ensure its healthy development and sustain the fertility of its parentage? These are questions I posed at the pharmaceutical laboratories of the Bristol-Meyers Company in Syracuse, New York, and in an essay in the *Syracuse Scholar* (Fall 1984). It was entitled "DNA Replication by Enzymes and Industry."

In my view, it is the basic biological sciences, interwoven with chemistry, that begot genetic engineering. The origin of the term was obscure to me when I recall first using it in a talk at Washington University in St. Louis in 1968. At the reception after the lecture, a friend admonished me that the term "genetic engineering" had a bad ring to it and urged that I find another. Some of that feeling persists, and so the term "biotechnology" has become a widely used euphemism for work that includes recombinant DNA studies.

Semantics can be significant. The term biotechnology can blur the important distinctions between biology and technology; genetic engineering can do the same by confusing genetics with engineering. As sciences, biology and genetics are responsible for the acquisition of basic knowledge. Technology and engineering, on the other hand, are responsible for the application of this knowledge to practical problems.

We recognize that science and technology are interdependent and often inextricably linked. We know that advances in science depend on techniques; the availability of techniques in turn depends on innovative and aggressive commercial development. When sophisticated instrumentation and fine biochemicals become commercially available and affordable, research is extended a thousand-fold. The explosive advances in cloning and DNA sequencing of the past few years would have been impossible without such developments as scintillation counters, centrifuges, commercial sources of enzymes, radioactive nucleotides, plastics, and the like.

While science is generally the pilot of technology and applied research, these roles have on occasion been reversed—science has been given a novel direction by the search for a practical solution. As mentioned earlier, Oswald Avery's concern with improving the treatment of lobar pneumonia impelled him to explore the nature of the pneumococcus to the point of discovering that DNA is the genetic material of life. Other fundamental medical discoveries have come from efforts to prevent and treat anemia, cancer, and heart disease.

So it is clear that technology can occasionally inspire or ease the way to a basic scientific discovery, but it must never be forgotten that technology would not exist without a foundation in science. This scientific base is often obscured or ignored when refinements in technology, coming one upon the other, make it seem that the marketed product is more important than the basic knowledge that created it.

Because biology and technology have become interdependent and even intertwined, it is tempting not only to lump them together but also to urge their amalgamation when they appear discrete. This amalgamation is encouraged so often that it is made to seem desirable and inevitable. Yet, there is a danger that unless carefully managed, the merging of biology and technology, either in the laboratory or in the funding of research, can weaken them both.

One hazard, a consequence of the increased popularity of applied biological research, lies in the lurid stories of possible misapplication of this new knowledge of genetic chemistry. We narrowly escaped federal legislation that would have confused and throttled both basic genetic research and its industrial application here and around the world.

In 1977, at the height of national concern about recombinant DNA, Senator Edward Kennedy sponsored a bill to regulate genetic research. When I pleaded that any such law was likely to be a serious mistake, he replied, "It is essential for the people to be in on the takeoff as well as the landing." By this he meant that the public should know ahead of time the consequences of research before permitting or supporting it.

I tried to explain to the senator that it was impossible to anticipate the consequences of basic research. It is common for people, including politicians, to cite the nuclear bomb as an illustration of science run amok. Let us ignore here the confusion between science, technology, and politics and ask: Could anyone have anticipated or regulated the chain of physical and mathematical research from Newton to Einstein and Bohr, from Heisenberg, Meitner, and Hahn to Fermi and Oppenheimer? At what point in the chain of basic discoveries in physics would one have intervened?

As for recombinant DNA, would anyone apprehensive about its development have forbidden Friedrich Miescher's chemical studies of the nucleus a hundred years ago? How could anyone have imagined that Avery's work on pneumonia would lead to the discovery that DNA is the substance of genes? Discoveries such as these provided the foundations that now make it possible for us to rearrange DNA, create genes, and thus readily manipulate heredity. If it seemed in any instance that recombinant DNA research were being misapplied, how could one suppress the simple and basic genetic and biochemical techniques that have become standard practice in thousands of medical science laboratories all over the world?

The analyses and rearrangements of DNA that form the drama of genetic engineering depend largely on a select cast of enzymes. Yet these actors were neither discovered nor selected to fill these roles. Some of the enzymes, uncovered in my laboratory, came from a curiosity about the mechanisms of DNA replication. In these explorations, sponsored by the National Institutes of Health and the National Science Foundation for over twenty-five years at a total cost of several million dollars, I neither anticipated nor promised any industrial application. Nor did any of my colleagues with comparable federally funded projects. In short, the genetic engineering industry now spread out before us sprang almost entirely from the pursuit of apparently irrelevant research in universities, made possible by the investment of many hundreds of millions of dollars by federal agencies over more than two decades.

If genetic engineering is to continue its healthy development and if the fertility of its parentage is to be sustained, it seems clear to me that burgeoning biotechnology enterprises must develop strong ties with academic science. The uneasiness in the relationships between academia and industry is understandable. The scientists, departments, and universities that provided the ideas and reagents, the techniques and machines, and the very practitioners of genetic chemistry and immunology are reluctant to be excluded from financial rewards they very much need. But there are obvious dangers if the university, as a nonprofit corporation, becomes entrepreneurial and employs its faculty and students for both academic and commercial objectives. There are major dangers, too, if biotechnology companies appropriate a generation of senior scientists as executives and consultants, junior scientists as employees, and then seal them off from the free exchange of new knowledge.

There can be no question about the desirability or even the urgency of putting this new knowledge to practical use. Nor is there any question that strong connections between academia and industry are needed to ensure that biotechnology develops well. Intimate relationships over many years between university chemistry departments and for-profit companies created our vast chemical industry; the same is true for physics and engineering departments and the electronics industry. The unusual features of the industrial application of biological science are that the basic technology came exclusively from academic laboratories and that it developed so rapidly. This is why university scientists and presidents have suddenly become concerned and vocal.

Related to this concern about academic-industrial relations in the development of industrial biology are the dangers of secrecy. Most scientists are convinced that secrecy in an academic setting impedes progress and profits no one. University and business leaders agree that corporate-sponsored academic research should be open. Well, sort of open. I would go much further. I believe that secrecy makes even less sense in an industrial setting than in an academic one.

An academic scientist in a highly competitive field can justifiably fear that the release of an idea, a hint that something works, or the source of a key reagent will enable another investigator to publish quickly and gain the priority for an important discovery. This appropriation of a new idea is far less likely in an industrial enterprise. Developing a major product, especially a pharmaceutical product requiring approval of the Food and Drug Administration, requires a commitment to spend a lot of money over a period of many years. For this reason, many factors other than having an original idea matter: the shrewd choice of a goal, high competence in attaining it, and effective marketing of the product.

The best insurance for industrial success is an open atmosphere that provides optimal access to all available information and advice. As significant advances are made, inventions can be protected each step along the way by a patent-applications team that interacts on the spot continually and knowledgeably with the scientists. In contrast, a company policy of secrecy closes the avenues of exchange with academic scientists and prevents useful collaborations with other companies. It is generally unappreciated that secrecy within the company itself shelters mediocrity and discourages the vigorous exchange that identifies excellence.

Scientific truths are logical. They can be stated simply and are hard to conceal. Technological developments, on the other hand, are more culturally bound and intricate. In fact, it may be hard to give them away. For these reasons, it is possible for Japan, with few contributions to basic science, to readily assimilate knowledge and excel in its application, while England, despite preeminence in basic science, has lagged behind in technological developments.

The argument against industrial secrecy is even more compelling in biotechnology than in most other areas because of the field's enormous potential for growth. The impact on medicine and industry of the new technical capacity to analyze, synthesize, and rearrange DNA and to exploit monoclonal antibodies will likely exceed even the most optimistic claims. Fluctuations in enthusiasm for biotechnology by venture capitalists simply reflect appraisals of how soon and by whom money will be made. While my bullishness for the power of this new technology to make practical advances in medicine and create new industries is strong, I am even more impressed by the impact of genetic engineering and the new immunology on basic biology.

A deeper understanding of the organization and functional control of chromosomes will soon emerge. And with that understanding will come interventions in aspects of growth, development, and aging that are now impossible. Within a few years, the most exciting prospects for medicine and industry will be subjects and products that no one has yet thought of. One may wonder which scientists will be sensitive to these new opportunities and what organizations will be equipped to seize them.

The currently active scientists who will lead these new advances have become bored with the routines of cloning and sequencing DNA and making hybridomas. Gifted young people entering biological science balk at joining the pack to clone hormones and vaccines. Instead, they seek new techniques to solve more challenging problems. In the past, pathfinders in biology have, with few exceptions, worked in academic institutions completely insulated from industry, and to a large extent this pattern might be expected to continue.

However, wise and farsighted management of a biotechnology enterprise might be able to change this pattern. Able scientists are interested in industry. Some are discouraged by the atmosphere often encountered in university departments: the emphasis on entrepreneurial skills of grantsmanship, the inevitable clashes with university bureaucracy, the obligation to serve on committees, the burden of heavy teaching loads, and the pressure to choose a safe, fashionable research program that will produce publications for the next grant application and academic promotion. In the face of these problems, one might see an industrial setting as offering several advantages: excellent resources, research objectives in interesting areas of science, fewer distractions, and a team spirit united for achievement.

One critical ingredient must be provided by industrial management if it wishes to capture and retain creative and productive scientists. It must provide an open atmosphere which encourages the scientist to discuss ideas, progress, and failures with colleagues in and out of his organization and to publish without restraint. Such an atmosphere is conducive to a flow of students, postdoctoral fellows, and visiting professors through the company. This is an atmosphere in which the scientist can feel confident that his creative ability will be fostered. It is a system which works when given a proper chance.

An urgent problem is how to nourish the goose of basic research that lays the golden eggs of industrial development. The astonishing advances in biological and medical science in the postwar period have proved that there is an ample reservoir of scientific talent for achievement that flourishes if provided with adequate resources and encouragement. The birth of genetic engineering and the revolutionary confluence of the medical sciences was made possible by the massive federal grants programs of the NIH and NSF in the postwar period. Periods of retrenchment by these agencies have blunted scientific progress. The budget for NIH in recent years has not provided adequate resources and grants for the highly trained and competent scientists who need them. There will simply be fewer research efforts, and those that are made, even by the most talented, will be less innovative and adventurous.

One often hears that private philanthropy and corporate contributions will fill the gap left by reduced federal support. This is utter nonsense. The gap in question is billions of dollars a year. Such private philanthropic resources are not available, nor are there equitable mechanisms for obtaining and distributing such sums except through federal taxation and administration. Will biotechnology companies, having obtained their science and scientists from the universities, share their fortunes with their academic parents and benefactors? To begin with, most of these fortunes are more fancied than real. And second, the logic behind sharing them is not all that persuasive. Would the universities wish to share the losses suffered by any of these bold corporate ventures should they fail?

Despite the inevitable irritation and envy about the unfairness of some of the financial windfalls, the recent commercial exploitation of basic molecular and cellular biology has accomplished several laudable things. It has made important products for medicine, industry, and agriculture. It has revitalized the pharmaceutical industry and is spawning related industries. It has created many attractive jobs in biology and genetics, where opportunities had become scarce. In doing so, it has provided an incentive for tuition-paying students to enroll in university programs of biological science that were beginning to languish. In addition, commercial success has raised the respectability of basic biological science among our fellow citizens and their governmental representatives, a stature that American dominance in the award of Nobel Prizes never did achieve.

But rich biotechnology companies are few in number; even fewer are able or inclined to give money away; and the rarest of them will not pour many millions of dollars into abstruse projects with gestation periods of twenty or more years. The only way to provide for the future of basic research is to preserve and enlarge the federal programs that support it. The massive basic research support by our federal government for the past thirty years has proved in the most compelling way that mechanisms are in hand to distribute large sums of money wisely and fairly, and that the talent is available to make astonishing progress with these resources. In practical terms, research, though expensive, is a bargain compared with the cost of disease.

Getting an adequate slice of the federal budget is a political problem. The electorate, Congress, and the administration must be convinced of the vital importance of investing in the training of scientists and in supporting the research of those who are highly competent and motivated. Lobbying, a key ingredient of the American political process, is essential. Unfortunately, such a lobbying effort is making a late start and must compete with lobbies that are keen, experienced, and resourceful and represent large and responsive constituencies.

Who will do the lobbying? A partnership between academia and the biotechnology and medical industries would be desirable. In urging such a collaboration between academia and industry for a massive and sustained lobbying effort in favor of basic support of science, the corporate and medical partners should be forewarned to expect rather little from the academics. Their attitudes and record of performance in this arena have been disappointing. For example, when you tell Professor Smith the shocking news that an esteemed colleague has not had his research grant renewed, he is likely to say, "Thank God, I still have two years to go on mine." A scientist warned of a fire down the hall may keep working until the flames reach his door.

Whereas societies of scientists do well at organizing meetings and publishing journals, their unwillingness and incapacity to enter the political process are traditional—perhaps congenital. In 1965, as President of the American Biochemical Society, I proposed an increase in the annual dues from $10 to $20 to fund a full-time person who would provide Congress, the administration, and the public with information about the nature and importance of biochemical research. Sources of such information, then and now, are fragmentary or even totally lacking. The Council of the society rejected my proposal, and I imagine the membership would have too. The ostrich-like behavior exhibited by biochemists is not confined to this scientific species.

It matters to all of us that basic science be vigorous and that the scientific base of medicine and industry be widened. We have the human and physical resources to do inspiring things. The cathedrals of science we build have no limits. There need be no jostling in designing and shaping these structures. The competitive challenge is for each of us to add beauty, body, and strength to these edifices of civilization and to endow them with a grandeur worthy of our lives.

By linking the physical, biological, and behavioral sciences through a common language, chemistry offers an understanding that creates a rational and aesthetic view of life. Despite the pressures to reap financial profits and the crusades to cure disease, the acquisition of basic, apparently irrelevant knowledge remains the lifeline of medicine.

Chapter 10

Reflections on My Life in Science

Circumstances appear to rule our lives, but the choices we make at many junctures determine just what the future circumstances will be. It matters too how quickly and decisively these choices are made. In the path I took from clinical medicine to enzymology, I made two major choices. Having been assigned to work on rat nutrition at the National Institutes of Health during World War II, I decided in less than a year to forget about a career in internal medicine in favor of working in a laboratory. Posing a discrete question and getting a straight and sometimes interesting answer after orderly effort was more gratifying to me than the unprogrammed hurly burly of medical practice. My second major choice was made after I had fed rats purified diets for three years. I had become frustrated with not knowing what vitamins really did, and so I abruptly quit doing animal nutrition to immerse myself in the new biochemistry. There I got to know enzymes for the first time and fell in love with them.

When asked whether a student with an interest in biochemical research should go to medical school, I do not recommend a circuitous course like the one I took from patients to rats to enzymes. Yet, these early experiences had some value. My medical training gave me an overview of the preclinical and clinical disciplines and, warranted or not, I have since enjoyed the confidence that I understood the practice of medicine and knew its limitations. I feel none of the uneasiness, common among my PhD colleagues, of having missed some rites or revelations essential for engaging in clinical research and the care of patients. I have felt more secure than most others in managing medical problems in my family, in participating in the policy decisions of a medical school, and in pursuing the conviction that medicine is best served by securing its foundations in science.

Deciding between a career in medicine and science has been and remains difficult for most people poised between those choices. Usually, the decision is put off for years and then some compromise arrangement is found, for a while. However, scientific research is too difficult to be done part-time. As with other creative endeavors, sustained productivity at a high level in science demands total devotion. I don't know of any exceptions.

In my nutritional studies with rats, which involved manipulating the environmental factors that influence their growth and maintenance, I found an agreeable transition from medicine to full-time laboratory research. After I became aware of the severe limitations of dietary studies with animals, I grew envious of those who were finding out so much more about vitamin functions by examining the nutrition of bacteria. When I heard about the enzymes whose catalytic functions explained why each of the vitamins was essential, I was even more eager to be among those making such discoveries.

Decades later, DNA and genes captured the spotlight from enzymes; but in my theater, enzymes kept the leading role. DNA and RNA provide the script, but the enzymes do the acting. For the cell, DNA is the construction manual, and RNA transcribes it into readable form, but the proteins, particularly the enzymes, carry out all the cellular functions and give the organism its shape.

I have been guided by the rule that all chemical reactions in the cell proceed through the catalysis and control of enzymes. Once, in a seminar on the enzymes that degrade orotic acid, I realized that my audience in the Washington University Chemistry Department was drifting away. Perhaps they had come to hear about "erotic acid." In a last-ditch attempt to capture their attention, I pronounced loudly that no cellular chemistry takes place without the action of an enzyme. At that point, Joseph Kennedy, the brilliant young chairman, awoke: "Are you trying to tell us that something as simple as the hydration of carbon dioxide (to form bicarbonate) needs an enzyme?" The Lord had delivered him into my hands. "Yes, Joe, there is an enzyme, called carbonic anhydrase. It enhances the rate of that reaction more than a million-fold."

Enzymes are awesome machines with a level of complexity that suits me. I feel ill at ease grappling with the operations of a cell, let alone those of a multicellular creature. I also feel inadequate in probing the fine chemistry of small molecules. Becoming familiar with the personality of an enzyme performing in a major synthetic pathway is just right. To gain this intimacy, the enzyme must first be purified to near homogeneity. For the separation of a protein species present as one-tenth or one-hundredth of one percent of the many thousands of other kinds in the cellular community, we need to devise and be guided by a quick and reliable assay of its catalytic activity.

No enzyme is purified to the point of absolute homogeneity. Even when other proteins constitute less than one percent of the purified protein and escape detection by our best methods, there are likely to be many millions of foreign molecules in a reaction mixture. Generally, such contaminants do not matter unless they are preponderantly of one kind and are highly active on one of the components being studied.

Only after the properties of the pure enzyme are known is it profitable to examine its behavior in a crude state. "Don't waste clean thinking on dirty enzymes" is sound dogma. I cannot recall a single instance in which I begrudged the time spent on the purification of an enzyme, whether it led to the clarification of a reaction pathway, to discovering new enzymes, to acquiring a unique analytical reagent, or led merely to greater expertise with purification procedures. So, purify, purify, purify.

I recall a lecture by James Watson at Stanford in 1969 in which he described to a large audience his discovery of a factor that directs the enzyme RNA polymerase where to start copying a message from the chromosome. He said at one point: "We did many experiments to find out how different cells use distinctive regions of their DNA. When we ran out of ingenious ideas, we did what Arthur Kornberg does. We purified the enzyme." I said to him later: "Jim, if you were real smart, you'd have done that in the first place."

Purifying an enzyme is rewarding all the way, from first starting to free it from the mob of proteins in a broken cell to having it finally in splendid isolation. It matters that, upon removing the enzyme from its snug cellular niche, you care about many inclemencies: high dilution in unfriendly solvents, contact with glass surfaces and harsh temperatures, and exposure to metals, oxygen, and untold other perils. Failures are often attributed to the fragility of the enzyme and its ready denaturability, whereas the blame should rest on the scientist for being more easily denatured. Like a parent concerned for a child's whereabouts and safety, I cannot leave the laboratory at night without knowing how much of the enzyme has been recovered in that day's procedure and how much of the contaminating proteins still remain.

With the purified enzyme, we learn about its catalytic activities and its responsiveness to regulatory molecules that raise or lower activity. Beyond the catalytic and regulatory aspects, enzymes have a social face that dictates crucial interactions with other enzymes, nucleic acids, and membrane surfaces. To gain a perspective on the enzyme's contributions to the cellular economy, we must also identify the factors that induce or repress the genes responsible for producing the enzyme. Tracking the enzymes that make the amino acid tryptophan, Charles Yanofsky, year after year, uncovers marvelous intricacies by which a bacterial cell gears enzyme production precisely to its fluctuating needs.

Popular interest now centers on understanding the growth and development of flies and worms, their cells and tissues. Many laboratories focus on the aberrations of cancer and hope that their studies will furnish insights into the normal patterns. Enormous efforts are also devoted to AIDS, both to the virus and its destructive action on the immune system. In these various studies, the effects of manipulating the cell's genome and the actions of viruses and agents are almost always monitored with intact cells and organisms. Rarely are attempts made to examine a stage in an overall process in a cell-free system. This reliance in current biological research on intact cells and organisms to fathom their chemistry is a modern version of the vitalism that befell Pasteur and that has permeated the attitudes of generations of biologists before and since.

It baffles me that the utterly simple and proven enzymologic approach to solving basic problems in metabolism is so commonly ignored. The precept that discrete substances and their interactions must be understood before more complex phenomena can be explained is rooted in the history of biochemistry and should by now be utterly commonsensical. Robert Koch, in identifying the causative agent of an infectious disease, taught us a century ago that we must first isolate the responsible microbe from all others. Organic chemists have known even longer that we must purify and crystallize a substance to prove its identity. More recently in history, the

FIGURE 10-1
A particular California auto license plate.

vitamin hunters found it futile to try to discover the metabolic and nutritional roles of vitamins without having isolated each in pure form. And so with enzymes it is only by purifying them that we can clearly identify each of the molecular machines responsible for a discrete metabolic operation. Convinced of this, one of my graduate students expressed it in a personalized license plate (Fig. 10-1).

Scientist, Teacher, Author, Chairman: In What Order?

I once shocked the Dean of Washington University School of Medicine by telling him that, as Chairman of the Department of Microbiology, my prime interest was to do research rather than teach. It has never been otherwise. Experiments are far more consuming and fulfilling for me than any form of teaching. Still, I have enjoyed a rather modest amount of formal lecture and laboratory instruction and have done it conscientiously. For the student, didactic teaching fails without the infusion of scientific skepticism and a fervor for new knowledge, and this comes naturally from someone dedicated to research. For me, some ten lectures a year freshen my awareness of basic subjects, and the preparation of a laboratory exercise has on one occasion opened a major avenue for my experimental work.

The most rewarding teaching for me has been in the intimate, daily contact with graduate and postdoctoral students. Well over a hundred of them spent from two to five years in my laboratory and were exposed to my tastes and obsession with the use of time. I felt closest to those who shared my devotion to enzymes and concern with the productive use of our most precious resource—each of the hours and days that so quickly stretch into the few years of a creative life. I recall in 1948 relating to Sidney Colowick and Ollie Lowry (both senior to me in age and experience) my failure with a certain procedure to purify a potato enzyme. "I wasted a whole afternoon trying that," I said. Colowick turned to Lowry and said with mock gravity, "Imagine, Ollie, he wasted a *whole* afternoon."

"An idea will not work unless you do," said Oswald Avery. The credo of hard work was professed by virtually everyone I know (from personal acquaintance or historical accounts) with successful careers in biology or biological chemistry. James Watson's *The Double Helix* (for which some have suggested the alternative title *Lucky Jim*) describes another course in which ingenuity and fortuity can in a single stroke bring fame and fortune.

At either extreme—speculating abstractly about complex phenomena or doggedly collecting data—success may come on occasion and draw acclaim. But the most consistent approach for acquiring a biochemical understanding of Nature lies in between. The novel is yet to be written that captures the creative and artistic essence of scientific discoveries and dispels images of the scientist as dreamer, walking in the woods awaiting a flash of insight, or as engineer, at an instrument panel executing a precisely planned experiment. Some intermediate ground—hard work, carefully planned, but with a touch of fantasy—is what I have sought for myself and my students.

If asked to rank the many varieties of mental torture, most scientists would place writing near the top of the list. As a result, scientific papers are usually put off or dashed off and demean the quality and value of the work they describe. Writing a paper is an integral part of the research and surely deserves the tiny fraction, say 5 percent, of the time spent finding something worth reporting. Yet, I confess to feeling uneasy when I see students and colleagues writing at their desks during working hours rather than busy at the laboratory bench. Perhaps taking time during the day to prepare a scientific report is unavoidable, but certainly writing a book seemed to me to be an unconscionable abdication from research—that is, until I wrote one.

In 1972 I gave a series of lectures on DNA replication in a graduate course at the City University of New York and then a comparable series, the Robbins Lectures, at Pomona College. I was dismayed at the uninformed and warped views that students and faculty had of the subject and my contributions, perhaps influenced by distorted articles and comments in *Nature New Biology* during the preceding years. Having tacitly agreed to publication of the Robbins Lectures, I thought it would be easy, with a little rearranging of extensive lecture notes and reprints, to write the book that would provide me with handy facts and references and also set the record straight for others.

Writing *DNA Synthesis*, a 400-page book which appeared two years later, surprised me in many ways. First, the effort was far greater than I imagined. Very little from the lecture notes and reprints could be lifted and placed in the right context and still remain readable. The required time had to be captured by exploiting a latent senile insomnia, generally from two to six in the morning; these became the most productive and enjoyable hours of my day. I was also surprised by the pleasure I found in reworking and polishing sentences and paragraphs for brevity and clarity, a satisfaction I had never found in crossword puzzles or other word games. Best of all, I could present my work, views, and excitement about the enzymology of DNA replication to an unexpectedly wide audience. The book—adopted as a text for some courses—became the reference source for writers of reviews of DNA replication and for authors of textbooks of biology and biochemistry.

By 1978, four years after *DNA Synthesis* was published, the book needed updating if it was to remain useful to me as well as to others. To get this started, Sylvy and I obtained an invitation to spend a month at the Rockefeller Foundation's Villa Serboloni in Bellagio overlooking the confluence of Lake Como and Lake Leccho. Favored by Roman emperors and still one of the loveliest spots on earth, the Villa provides ten resident scholars a place to live and write in eighteenth-century elegance. Without keys, car, money, or telephone, and with superb food and other services, this idyllic retreat is competed for by humanists from everywhere but remains virtually unknown to scientists. The weekly Villa bulletin that listed the mission of my stay to be the revision of a four-year-old book was beyond the comprehension of my colleagues, who regarded chronicles of nineteenth-century history as mere journalism. They assumed I had botched the first edition very badly.

DNA Replication came out in 1980. Twice the size of its predecessor, it was really a new book in scope and organization as well as in expanded contents. Despite its inflated size, it was a better book and found a wider readership. However, progress in this field is so rapid that revisions are needed annually. As an experiment in publishing, I assembled a 273-page *1982 Supplement to DNA Replication*, which extended the life (and sales) of its parent. The publishers objected to the "1982" in the title and correctly saw it as an advertisement of obsolescence. There have been no further biennial supplements.

If teaching and book writing are regarded as deviant activities for a dedicated scientist, then surely the administrative work of a departmental chairman should be beyond the pale. Yet, I served as chairman for over twenty years and never found it to be a serious intrusion on my time or attention. On the contrary, the benefits of creating and maintaining a collegial and stimulating scientific circle were well worth the investment I made. With excellent administrative assistance and the eager participation of my faculty colleagues, direction of departmental activities took no more time than being a conscientious member of the department.

Involvement in broader medical school and university affairs is a far different matter. I never found the skills and patience to function at these levels. For me, the most burdensome feature of being a departmental chairman was obligated service on the Executive Committee of the Medical School, which was preoccupied with budgets, promotions, interdepartmental feuds, and salaries. In six years at Washington University and ten at Stanford, I cannot recall a deliberate discussion of science or educational policy. No wonder I had no interest in being the dean of a medical or graduate school on occasions when this possibility was raised.

Increasingly conspicuous in current scientific life are the extramural administrative and educational activities, which, with the attendant travel, may consume half the time of prominent members of a science faculty. Lectures and visiting professorships, scientific meetings and society councils, government panels and advisory boards, consultantships in industry—all are prestigious, diverting, less demanding than research, and terribly tempting. I have done less than most but have been unable to resist some participation, particularly in writing and testifying for federal support of research and training. Most recently I have also been involved in the founding and development of a biotechnology enterprise (the DNAX Research Institute, Inc., later acquired by the Schering-Plough Corporation), whose mission is to apply the techniques of molecular and cellular biology to the therapy of diseases of the immune system.

All these nonresearch activities, in and out of the university, fail to give me a deep sense of personal achievement. In research, it is up to me to select a corner of the giant jigsaw puzzle of nature and then find and fit some missing pieces. When, after false starts and fumbling, these pieces fall into place and provide clues for more, I take pleasure in having done something creative. By contrast, in my other activities, which are just as personal, all I do, it seems, is try to use common sense and behave in a fair and responsible way, as anyone else would. This personal sense of creativity is the reason why research has been so dominant over my teaching, writing, and administrative activities (in sharply descending order of importance to me), but I sometimes wonder whether, based on their value to science, this order should be reversed.

Consider the creation and administration of the Stanford Biochemistry Department. In polls of peers, the department was accorded the top rating for many years and is regarded as a major source of discoveries basic to recombinant DNA and the genetic engineering revolution. Over five hundred people trained in the department now staff and direct departments of biochemistry and molecular biology all over this country and around the world. The organization, development, and preservation of this notable faculty against strong centrifugal and attractive forces, I cannot deny, has been a unique achievement.

As for writing, the textbooks on DNA replication, with over forty thousand copies sold, have made it easier for others to enter and work in this field. Fred Sanger, who won a second Nobel Prize for his technique of determining the sequence of a DNA chain, told me that the idea of using a dideoxy analog of deoxynucleoside triphosphates came from reading these books. More than providing a readable account of a forbiddingly specialized area of biochemistry, the books have helped revive an appreciation that enzymology offers a direct route toward solving biological problems and is the source of reagents for the analysis and synthesis of a great variety of compounds for all branches of biological science.

With regard to teaching, a mentor's assumption of credit for the success of a student has always puzzled me. There simply are no controls in these experiments. How do I know, given a motivated, gifted student, whether I have been a help or hindrance? Nevertheless, having involved myself in the daily scientific lives of my students, I may have guided some of them in directions that attract me and thereby diverted them from a career in medicine or pure chemistry to the love and pursuit of biochemistry and enzymes. These progeny now include illustrious figures in science who have spread this gospel to a widening circle of grandstudents and great-grandstudents.

If the case can be made that my activities in administration, writing, and teaching have made a unique contribution, then certainly a further case can be made that my discoveries in science have not. Very likely, they would have been made by others soon after. After all, Darwin's epochal discovery of evolution through natural selection was made simultaneously by Alfred Wallace, continents away, and the Watson-Crick structure of DNA would have come to Rosalind Franklin, Maurice Wilkins, or Linus Pauling within a year. Yet in the last analysis, I will argue that for me it was the research that mattered most, because all of these other activities and the attitudes I brought to them were shaped by it.

Some Shadows on a Sunny Scene

Science, as a creative activity, is an art form. In trying to expose a hidden view of Nature, the researcher uses or fashions tools to dig deeper or to enlarge a perimeter. As with other creative activities, the experiments generally fail and can be rather discouraging.

These days, when I no longer do experiments with my own hands, I marvel at the patience and fortitude of my students—and cannot easily recall my own—in the face of repeated failures. There are the annoyances of spoiling an experiment by a clumsy movement, omitting an essential reagent, mixing up the order of test tubes, or neglecting a basic rule of chemistry or biology. Then there are the frustrations of tracking an interesting lead only to rediscover something familiar.

The worst kind of failure, and a common one, is the inability to repeat what appeared to be a novel finding enlarged by fantasy to a great discovery. Why the worst? The first experiment may have given the hint of working because of a fortunate catenation of preparing an extract in a particular way from cells at a certain stage, adding reagents in a unique sequence, incubating the mixture at a selected temperature for a given length of time, and so on—all at the ragged edge of failure. The temptation is great to go on gambling with different combinations of the numerous parameters in order to repeat that first experiment. And why is this failure so common? Simply because one doesn't repeat the experiment that shows no glimmer of interest.

The occasional success in wresting a secret from Nature fuels the will to go on. In addition to the satisfaction of making a discovery is the approbation we get from others. Unfortunately, the measure of recognition often depends on skillful reporting and advertising and on the luck of being in the roving spotlight, even more than it does on the importance of the discovery. One of the gentlest and kindest scientists I have known once told me he wished that publication of results were anonymous. Yet, years later, after he had pioneered a new area of research, he complained that others who had entered the field later were getting undue credit.

The publish-or-perish rule governs all of us who do research and must rely on grants to support it. Timely and numerous publications matter, as does grantsmanship—skill and shrewdness in preparing a grant application. A well-selected title, a popular subject, and voluminous detail help in guiding the application to the best-endowed agency (such as the National Institutes of Aging and Cancer) and the most sympathetic refereeing group. With the tightening of research budgets in recent years, there have been casualties among qualified scientists at the same time that less competent ones were obtaining expanded support.

Despite these inequities, the support of science is incalculably better than it was when I entered science. In 1942 research positions were rare, and there were no grants worth mentioning. The select few research workers, like most artists, could afford only the simplest materials and had to subsist on meager salaries. Today, the motivation to do science is not tested by privation, and the opportunity to do creative work in a well-compensated, prestigious position is open to large numbers of people with training and intelligence. Unlike many, I feel keenly my good fortune to have been supported in a career in science limited only by my own drive and ability.

The natural regrets of not having done more in less time have been balanced by my optimistic outlook and the deep aesthetic pleasure of a glimpse now and then into an unanticipated, awesome beauty of nature. Over a lifetime, I have watched a vastly expanding picture of the living form with the added satisfaction of having helped in uncovering a few of its features. And I have had nearly lifelong associations with scientists whose friendship I trust, whose company I enjoy, and whose creative work has given me almost as much pleasure as my own.

My only disappointments have been in the occasional displays of human frailties in some of those who do science and administrate it. What makes science unique is the discipline, rather than the practitioner. Verifiability and incremental progress distinguish science from all other art forms. The scientist must be scrupulous in any claim of progress, because he bears a burden of guilt until confirmation by others and an enlarging significance of the discovery have proven him innocent. His personal behavior may be another matter. The revelations of some scientists' unseemly behavior made Watson's story of discovery in *The Double Helix* fascinating to readers who lacked an awareness of the human nature of scientific activity.

Scientists generally make poor administrators. Self-selected for a greater interest in "things" (such as molecules and cells) than in people and their social interactions, the scientist usually does not do well as chairman of a department or dean of a school. He frequently becomes overly concerned with procedural details, displays an intolerance of idiosyncratic styles, and fails to exercise scientific leadership. The deanship of a medical school is a virtually impossible job, and the wisdom of anyone who undertakes it is immediately suspect. As the only professional school in the university fully engaged (through its affiliated hospitals) in a big business activity, the medical school is beset by nettlesome community relations, hostile practicing physicians, outlandish budgets, Medicare regulations, and a huge faculty. Still, there is no excuse for a swollen bureaucracy and alienation from the school's fundamental mission: the acquisition of new knowledge. Fortunately, grants for research from federal agencies, the lifeblood of discovery, are made directly to the scientist and are thereby insulated from mischievous departmental and school politics.

My most bitter experience in academia was in 1969 in the heat of the Vietnam war and the civil rights struggle. To end the debacle abroad and to correct racial injustice at home, faculty communities across the country reacted with moblike hysteria and abandoned all regard for academic freedom and standards. Sidney Hook gives a graphic description of the events at New York University in a chapter entitled "Walpurgisnacht" (night of the witches' Sabbath) in his autobiography *Out of Step;* he also notes the remarkable absence of a proper historical account of this frightening period in academic life.

At Stanford University, the research and teaching laboratories of the Electrical Engineering Department were occupied and disrupted by students and others, with the encouragement of some faculty members. I served at the time on a small presidential crisis committee and was appalled that this outrage drew no broad response from the faculty or effective action from the administration.

Another incident was more personal and painful. I had been critical at a faculty meeting of what I regarded as excessive indulgence to a medical student who had repeatedly failed his courses. Because he was identified as racially black—not at all obvious to me and irrelevant if it were—I was accused of racism in the student newspaper and physically threatened by the black students' organization. Instead of rallying to support me, some of my faculty colleagues whispered their disapproval of my "insensitivity." How could *they* have been so insensitive? They must have known of my own struggles with prejudice and my long record of compassion with victims of social injustice and devotion to liberal causes. In the mood of the moment, they abandoned their allegiance to scholarship in their passion for righting an ancient and ugly wrong.

The consequences of this general erosion of academic values were predictable. A student-dominated committee on admissions, directed to see that 20 percent of the class were of black and Hispanic origins, went farther and discriminated against minority applicants whose good scholastic qualifications could be attributed to an "advantaged" background. Under these pressures, grades were eliminated and difficult courses were made elective. These attitudes permeated the student body and favored appointments of faculty with a similar lack of scholarly devotion. After a decade of this "cultural revolution" came a gradual return to the sane recognition that scientific rigor is the ultimate basis of medical progress, good clinical practice, and physicianly responsibility.

The Virus of Anti-Semitism

Some viruses are eradicable. The virus of smallpox has been removed from the population and survives only in a few well-guarded museums. Similar conquests may some day be achieved with polio and other pathogenic viruses. Sadly, the same success is unlikely with the more devastating agent of anti-Semitism, epidemic for twenty centuries and virulent around me early in my career.

In the grade schools and high schools of Brooklyn, I was enclosed in a circle of Jewish students and friends and was unaware of any anti-Semitism directed at me. This innocence persisted until my senior year at the academically prestigious City College of New York, whose student body was then 90 percent or more Jewish. Then came the disappointment of being rejected by virtually all of the many medical schools to which I applied. But it came as no surprise. All my classmates, most with superb records, were also rejected. I resented then that at the College of Physicians and Surgeons of Columbia University, a close neighbor of City College, an endowed scholarship for a City College graduate went begging for nine years because there were no candidates. To this day it rankles me.

Finally, I was admitted to the University of Rochester Medical School, one of its quota of two Jews in an entering class of forty-four. For a while the pain of the earlier rejections was erased; also blurred was the awareness that in 1937 all but five of two hundred premed classmates at City College had been denied entrance to medical schools.

My first shock at Rochester was hearing anti-Semitic remarks and realizing how much more numerous they must be in my absence. The worst was being denied academic awards and research opportunities because I was Jewish. I particularly coveted the pathology fellowship—a year's research and special training awarded to two students at the end of the course in pathology. The fellows were chosen by George H. Whipple—Nobel laureate, chairman of the department, dean of the school, and God of the Rochester medical universe. The anointed selectees moved on an inside track to the best internships and careers in academic medicine. I knew that my class performance was better than that of the students awarded the fellowships, and I later learned that I had ranked first in the class. There were less attractive fellowships in other departments, but none was offered me.

Sixteen years later, in my capacity as Chairman of the Department of Microbiology at Washington University School of Medicine, I attended a Pathology-Microbiology Teaching Conference. In a group discussion of how to evaluate student performance, I clashed with Sidney Madden, who had been my instructor in pathology at Rochester and was now the department chairman at the University of California in Los Angeles. After the session, he asked me why I had been so insistent on strict objectivity in grading. "Arthur, you know how much emphasis the Dean (Whipple) placed on the character of the student," he said. "Look, Sid," I responded, "I won't beat about the bush. Whipple was anti-Semitic and his judgments were colored by it." Madden reflected for a moment and countered: "That's true, Arthur, but the Dean didn't like Italians either."

I have wondered all these years how my medical school professors, all otherwise decent and intelligent men, could have tolerated and participated in this discrimination against me. Assuming the absence of a personal dislike or some visceral reaction against Jews, they might have thought it a kindness to keep me from a path in which I would find the doors closed along the way. Or else prudence might have persuaded them, no matter how unfair, not to waste an opportunity on someone who would not be given the chance to use it. Yet, there was one professor at Rochester who stood out. William S. McCann, Chairman of the Department of Medicine, the only one to appoint a Jewish Chief Resident, did everything he could to help me. He persuaded a wealthy patient to endow a scholarship of which I was the first recipient, found one hundred dollars (a significant sum in those days) for me to buy bilirubin to pursue my first research project (on jaundice), appointed me to an internship in medicine and then to an assistant residency, and was likely grooming me for a career in academic medicine.

About twenty years after leaving medical school, in the mid-1960s, I was visited at Stanford by W. Allen Wallis, then President of the University of Rochester, who wanted to interest me in becoming Dean of the Medical School. When I made it plain that I had no ambition for that level of administration, he asked whether my sentimental attachment to Rochester might make a difference. I then described the residual bitterness I felt toward Rochester for the treatment I had been given. "I was here at Stanford at that very time," Wallis said. "Conditions were just as bad as at Rochester." Of course he was right. Anti-Semitism was virulent throughout academia before World War II.

In *Joining the Club: A History of Jews and Yale* (Yale University Press), Dan A. Oren presents a well-documented account of the restriction of enrollment of Jews and a description of the club-like atmosphere which rejected diversity and snubbed scholarship. As late as 1950, "of the twenty-four professors on the Yale medical school board of permanent officers not one was a Jew. Professor Joseph Fruton, a biochemist, was the sole Jew among the twenty-seven full professors on the teaching faculty." Emil Smith, Professor Emeritus of Biochemistry at UCLA, recalls that those who voted to appoint Fruton to the position of full professor did not know at the time that he was Jewish.

At the start of this century even Albert Einstein had problems getting his first faculty appointment. As described by Abraham Pais in his biography "*Subtle is the Lord . . . ,*" the proposal of Einstein's election to the faculty at the University of Zurich was made by one Alfred Kleiner, who wrote: "Today Einstein parades among the most important theoretical physicists and has been recognized rather generally as such since his work on the relativity principle . . . uncommonly sharp conception and pursuit of ideas . . . clarity and precision of style." The faculty response in its final report read: "These expressions of our colleague Kleiner, based on several years of personal contact, were all the more valuable for the committee as well as for the faculty as a whole, since Herr Dr. Einstein is an Israelite and since precisely to the Israelites among scholars are ascribed (in numerous cases not without cause) all kinds of unpleasant peculiarities of character, such as intrusiveness, impudence, and a shopkeeper's mentality in perception of their academic position. It should be said, however, that also among the Israelites there exist men who do not exhibit a trace of these disagreeable qualities and that it is not proper, therefore, to disqualify a man only because he happens to be a Jew. Indeed, one occasionally finds people also among non-Jewish scholars who in regard to a commercial perception and utilization of their academic profession develop qualities which are usually considered as specifically 'Jewish.' " The faculty voted in March 1901 to appoint Einstein: ten were in favor, one abstained.

Despite the prevalence of anti-Semitism in academia in Western society, there have been pockets of tolerance and individuals who bravely rejected discrimination. Prominent among them were F. Gowland Hopkins at the University of Cambridge, Hans T. Clarke at Columbia University, and Carl Cori at Washington University. In welcoming scientists of all nationalities and providing a haven for refugees from oppression, they developed the foremost departments of biochemistry of their eras. By contrast, the vicious anti-Semitism and anti-intellectualism in Germany during the 1930s drove its science from the apex to a lowly status from which it is only beginning to recover a half-century later.

When I moved to St. Louis in 1953, the bars against Jews were beginning to lift; the segregation of black people, however, was still severe. My predecessor as chairman of the department was Jacques Bronfenbrenner. Because he was Jewish, he had felt constrained to limit the number of Jews in his department to avoid any appearance of favoritism. I was determined to be racially blind. In the thirty-five years at Washington University and Stanford, race was never thought of, let alone considered, in our choice of faculty, staff, or students. In many cases, I never did learn whether someone was Jewish or not. Yet in view of the high density of Jews in biochemistry, I had to wonder about the racial attitudes of those biochemistry departments which, over several decades, had few or no Jewish students or faculty.

The current academic scene bears little resemblance to the one I knew early in my career. Jews populate all areas of scholarship and levels of university administration. My three sons and my students find it difficult to believe it was ever different. They have been threatened neither by the virus of poliomyelitis nor by that of anti-Semitism. There is of course a profound difference between these two afflictions. Suppression of the polio virus depends on the rational development of a vaccine. Should this immunity be overcome by a viral mutation, the setback, while serious, would be counteracted by development of a new vaccine. There can be no vaccine for anti-Semitism. I cannot believe that this disease, pandemic for many centuries and prevalent here until so recently, will not break out again.

Basic Research, the Lifeline of Medicine

Penicillin and other antibiotics are the most dramatic therapeutic advance in medicine in my lifetime. When I was a medical student and intern before the advent of antibiotics, the treatment of lobar pneumonia was discouragingly ineffective. One out of four patients died. Subacute bacterial endocarditis was invariably fatal. Rheumatic fever and acute nephritis were prevalent.

Many people know that antibiotic therapy was not discovered at the bedside nor even in a clinical pharmacology laboratory. Bullrings in Spain have statues of Alexander Fleming, who in 1929 noticed the inhibition of bacterial growth around a penicillium mold contaminating his Petri plate. The apotheosis of Fleming by gored bullfighters is exaggerated because the practical use of the penicillium mold was not discovered by Fleming. It took Ernst Chain, a biochemist, and Howard Florey, a pathologist, to apply this knowledge ten years later to isolate penicillin and demonstrate its clinical utility.

There is much more to the penicillin story than that. Basic inquiries and findings essential to Fleming's discovery started at least fifty years earlier. Fleming would never have made his observation without the agar Petri plate. But, more important, he would never have understood what he saw on the plate were it not for the firm foundations of bacteriology and immunology.

One might assume that Chain and Florey undertook the isolation of penicillin because of its possible clinical potential. Not at all. With the encouragement of Florey, Chain started his research on penicillin only because he was curious about the dissolution of bacterial walls by enzymes such as lysozyme. He thought penicillin was an enzyme too and wanted to understand the mechanism of its action. He was surprised to find that penicillin was a molecule small enough to pass readily through the pores of a dialysis membrane which stopped passage of molecules as large as enzymes. Penicillin was not an enzyme at all. This discovery immediately presented the possibility that penicillin, as a low-molecular-weight compound, could be administered as a drug to animals. With the technique of freeze-drying, which had just become available, Chain was able to concentrate and preserve penicillin and then to prove its therapeutic efficacy.

Why were Chain and Florey so quick to test their very crude penicillin preparations for clinical efficacy? Their prompt decision was likely conditioned by the discovery in the mid-1930s that sulfonamides inhibit microbes and yet are not toxic to animals. Thus, an agent could selectively interrupt growth of microbes without affecting the animal host, and this observation gave Chain and Florey the confidence to test their penicillin preparations in infected mice. It may also explain why Fleming, ten years earlier, believing that only the host immune system could combat infections, regarded chemotherapy as implausible and so failed to test his crude penicillin preparations for therapeutic value.

I choose penicillin as an example of the importance of basic research because its history is so recent and dramatic. A generation earlier, I would have cited x-rays. They were discovered by Wilhelm Roentgen in 1895 and immediately applied to medicine. X-rays were not discovered because such a technique was needed in medicine and surgery. X-rays were discovered because physicists were curious about an utterly esoteric question: how electricity behaved in a vacuum.

If we examine virtually any drug or procedure of proved efficacy in medicine, the history of its development is essentially the same. The pathway of a clinically relevant discovery is a complex sequence of many steps and branches from many disciplines. We find that the early and major part of the pathway of discovery is generally unrelated to any specific clinical objective.

When we extoll what basic research has done for us lately, we are simply reflecting on the achievements that inevitably accrue from the practice of science. The goals and attitudes in research have not changed appreciably for hundreds of years. The practice of science depends on the same familiar human qualities required in other professions, in art, and in business. What distinguishes the practice of science from the practice of medicine, law, and politics is not the scientist but the discipline he must follow.

It is the essence of scientific discipline to ask small, humble, and answerable questions. Instead of reaching for the whole truth, the scientist examines small, defined, clearly separable phenomena. The pattern of science is a step-wise extension of what came before. Whereas the doctor must treat the whole patient, and at once, the scientist can isolate the smallest facet that intrigues him and grapple with it for as long as it takes.

I was surprised that these verities presented in a lecture to an elite group of clinical researchers at their annual meeting in Carmel, California, in 1976 drew a warm response. It seems that scientists, as well as laymen, need to be reminded about the nature of scientific discovery and the hazards of seeking easy solutions to complex problems.

I began by describing an amusing thought I had while listening to a seminar about a mutant of *E. coli* that was defective in its ability to regulate cell division. Because of its defect, this particular mutant produced minicells totally lacking DNA. Nevertheless, the minicells respired and even synthesized proteins to a limited extent. As someone with a vested interest in DNA, I became increasingly worried, as the seminar unfolded, that these cells could manage so well without DNA. Surely, if these cells had a future without DNA, then my career interest in DNA did not.

Of course, such minicells have no reproductive future, and their intelligence for making new proteins is severely limited. Yet minicells, like enucleated eukaryotic cells (red blood cells, for example), do have a present. Their membranes give them the integrity to retain precious macromolecules and permit them to engage in a lively and profitable exchange with the nutrients in their environment. By contrast, the massive intelligence represented by DNA has by itself neither a future nor a present. DNA without a cell to sustain and express it has no physiologic meaning. Thinking about this *E. coli* mutant made me more aware of the relative values of ideas and intelligence on the one hand and of their expression on the other.

Before we can know the true value of an idea, it has to be expressed or implemented. It must be marketed, used, tested. This is as it should be. Unfortunately, the reverse is not true. Marketing can be successful, for long periods, even though it lacks a sound idea or any substance. This domination by marketing is true of all aspects of human behavior, including the practice of medicine and science.

The marketplace of medicine has many popular products and practices that are harmful, useless, or of uncertain value because they lack firm foundations in scientific understanding. This, it seems to me, has been and remains the key problem of medicine. The search for essential, basic knowledge is tedious, difficult, and time-consuming, whereas the appeal of quick rewards is tough to resist. We are disheartened and diverted by the easy and quick marketing successes of practices that lack this essential, basic knowledge.

I have often wondered what would have been my fate had World War II not shunted me from a career in clinical medicine. I, like my audience in Carmel, would likely have found some academic niche in clinically relevant research. I admire those who do this job well, for I fear I would not have been suited to it. One must carry on enough clinical practice to be sufficiently sharp to teach clinical medicine, to do justice to the patient's needs, and, especially in recent years, to earn money. But attending to patients is time-consuming and often distracting from sustained laboratory research. On top of that, clinical research is exceedingly difficult, and the intellectual rewards are hard won and too often fragmentary.

Another problem is the choice of what research to do. For the clinical investigator there is the temptation to engage in many research projects because of stimulation by so many intriguing clinical problems. I can think of nothing more destructive of the productivity of the clinical investigator than a failure to focus sharply on a single problem over a long space of time. The chief cause of this failure to advance knowledge is not a lack of effort, or motivation, or opportunity. It is not a lack of creativity, or intelligence, or training. The clinical investigator fails when he lets problems choose and dominate him, rather than the reverse. The investigator must ask a small and modest question, focus on it in laser-beam fashion, and then maintain the focus until the beam burns through. The most creative and productive work will still be done by such an individual or a small group rather than by a large team.

I hear it said that the primary function of a medical school is to teach medical students. How tragically sterile an outlook that is! Of course, the medical school must, like other technical schools within the university, prepare students with the basic knowledge, skills, and experience to become fully qualified practitioners. However, the central mission of a medical school is to advance the knowledge of medicine and to imbue students with this spirit. By asking probing and penetrating questions that challenge a current dictum, we train physicians who will remain attentive to new knowledge and to whom we would entrust the care of our families. I also believe that a great deal of teaching is done to satisfy the ego and status of the teacher and department rather than for the needs of the serious student. As for the proliferation of university administrative work, it is unpardonable. Much of the time spent on grant and fellowship applications, committees, surveys, and the rest distorts and destroys the primary responsibility to advance the knowledge base of medicine.

I have also wondered on occasion what I would do were I engaged in the full-time, demanding practice of medicine. I put myself in the place of the practitioner who must rely on his own hands and wits to treat patients, to earn his living, and to support his malpractice insurance policy. Isn't he doing enough, just by keeping abreast of new developments in medical practice? Is it reasonable to expect him to advance knowledge, too? I believe it is. The essence of professionalism, whether in art, literature, or medicine, is to solve problems that provoke intellectual curiosity and to be creative. Only thereby does the physician perform as an artist rather than a practitioner.

I know that, were I in clinical practice, I would become uneasy and curious about many of the puzzling questions that physicians constantly encounter. I would try to select just one of these many questions. It would be a question about which I could most easily collect quantitative data from my patients. I would collect the data, sort and analyze them, and then collect more data. The discipline of this exercise would sharpen my practical skills as a physician. But more important, after studious collection of data about a specific question for many years, I would surely know a little more, if ever so little more, that would advance medical knowledge. And unless medical knowledge advances, however incrementally, it will surely regress, because the status quo is continually challenged with new problems—novel toxins, resistant organisms, and unsubstantiated myths about health and disease. Experience has taught us that research is the lifeline of medicine. What a wonderful difference it would make if tens of thousands of physicians could report once in their lifetime how they had reshaped a fact of medical science!

Chronology

Bibliography

Glossary

Index

Chronology

1918   Born in Brooklyn, N.Y., to Joseph Aaron and Lena Rachel (née Katz); brother, Martin, age 13; sister, Ella, age 9.

1933   Completed Abraham Lincoln High School (Brooklyn); entered City College of New York (CCNY).

1937   Bachelor of Science degree; entered University of Rochester School of Medicine, Rochester, N.Y.

1941   M.D.; started internship in internal medicine, Strong Memorial Hospital, N.Y.

1942   Completed internship; entered U.S. Public Health Service assigned to U.S. Coast Guard (U.S. Navy); sea duty in the Caribbean as ship's doctor; transferred to Nutrition Section, National Institute of Health (NIH).

1943   Married Sylvy Ruth Levy.

1946   Research training with Severo Ochoa at New York University School of Medicine.

1947   Research training with Carl Cori at Washington University School of Medicine, St. Louis; son Roger born; returned to NIH to start the Enzyme Section.

1948–50 Sons Tom and Ken born.

1953   Resigned from NIH (rank of Medical Director) to become professor and chairman of the Department of Microbiology, Washington University.

1959   Moved to Stanford University School of Medicine as professor and chairman of the Department of Biochemistry; Nobel Prize in Medicine or Physiology, shared with Severo Ochoa.

1986   Sylvy died after a long illness.

1988   Married Charlene Walsh Levering.

Bibliography

Chapters 1 and 2

Ames, B. N. 1983. Dietary carcinogens and anticarcinogens. *Science* 221:1256–1264. The cover of this issue carried—in block letters—"Eat and Die."

Carpenter, K. J. 1986. *History of Scurvy and Vitamin C.* Cambridge: Cambridge University Press. A well-researched account of this fascinating disease and its solution.

Fruton, Joseph S. 1972. *Molecules and Life: Historical Essays on the Interplay of Chemistry and Biology.* New York: Wiley-Interscience. A scholarly history of fermentation, enzymes, energy metabolism, and DNA.

Holmes, F. L. 1985. *Lavoisier and the Chemistry of Life: An Exploration of Scientific Creativity.* Madison: University of Wisconsin Press.

Kornberg, A. 1942. Latent liver disease in persons recovered from catarrhal jaundice and in otherwise normal medical students as revealed by the bilirubin excretion test. *Journal of Clinical Investigation* 21:299. My first publication; includes me as one of the subjects.

Kornberg, A., F. S. Daft, and W. H. Sebrell. 1942. Production and treatment of granulocytopenia and anemia in rats fed sulfonamides in purified diets. *Science* 98:20–22. My first report of full-time research.

——— 1944. Mechanism of production of vitamin K deficiency in rats by sulfonamides. *Journal of Biological Chemistry* 155:193–200. My first publication in the most prestigious journal of biochemistry.

Schlenk, F. 1985. Early research on fermentation—a story of missed opportunities. *Trends in Biochemical Sciences* 10:252–254.

von Haller, A. 1962. *The Vitamin Hunters.* Philadelphia: Chilton Co. Useful but hardly a match for *Microbe Hunters* (see below).

Chapters 3 and 4

Davis, Bernard D., et al. 1980. *Microbiology.* 3rd ed. New York: Harper & Row, Inc. An excellent general textbook.

De Kruif, Paul. 1926. *Microbe Hunters.* New York: Harcourt, Brace. Dramatic accounts of the early explorations of microbial diseases.

Dobell, Clifford. 1960. *Anthony van Leewenhoeck and His "Little Animals."* New York: Dover. The best account of the fascinating man who built the first microscopes, with which he discovered an entirely new world of microbes and tiny things.

Dubos, René J. 1950. *Louis Pasteur, Free Lance of Science.* Boston: Little,

Brown and Co. A lively and insightful appreciation of the man and his colossal achievements.

Roueché, Berton. 1967. *Annals of Epidemiology and Other Collections of Stories from the New Yorker.* Boston: Little, Brown and Co. His enthralling articles continue to appear in the *New Yorker.*

Stanbury, J. B., et al. 1983. *The Metabolic Basis of Inherited Disease.* 5th ed. New York: McGraw-Hill. Comprehensive textbook of the scientific bases of the numerous genetic disorders.

Chapters 5 and 6

Goulian, M., A. Kornberg, and R. L. Sinsheimer. 1967. Enzymatic synthesis of DNA. XXIV. Synthesis of infectious phage ϕX174 DNA. *Proceedings of the National Academy of Sciences USA* 58:2321–2328. Experimental description of "creation of life in the test tube."

Kornberg, A. 1960. Biologic synthesis of DNA. *Science* 131:1503–1508. The Nobel Prize lecture delivered in Stockholm in December 1959.

McCarty, Maclyn. 1985. *The Transforming Principle.* New York: Norton. An authentic history by one of the participants in the discovery that DNA is the genetic material.

Chapters 7 and 8

Bramhill, D., and A. Kornberg. 1988. Duplex opening by dnaA protein at novel sequences in initiation of replication at the origin of the *E. coli* chromosome. *Cell* 52:743–755. Contains pertinent references to the recent literature.

Kornberg, A. 1988. DNA replication: a minireview. *Journal of Biological Chemistry* 263:1–4.

Chapters 9 and 10

Crick, Francis. 1988. *What Mad Pursuit.* New York: Basic Books. The "personal view of scientific discovery" of the foremost theoretical molecular biologist.

Davis, B. D. 1986. *Storm over Biology: Essays on Science, Sentiment and Public Policy.* Buffalo: Prometheus. Reviewed by me in *Nature* 324:172 (1986).

Hall, S. S. 1987. *Invisible Frontiers: The Race to Synthesize a Human Gene.* New York: Atlantic Monthly Press. A lively account of the contest between two biotechnology companies to patent the cloning of the gene for insulin.

Kornberg, A. 1976. Research, the lifeline of medicine. *New England Journal of Medicine* 294:1212–1216. Adapted from a lecture presented on February 5, 1976, in Carmel, California, to the Western Section of the American Federation for Clinical Research and the Western Society for Clinical Research.

——— 1984. DNA replication by enzymes and industry. *Syracuse Scholar,* fall, pp. 49–56.

——— 1987. The two cultures: chemistry and biology. *Biochemistry* 26:6888–6891. Adapted from a lecture at a meeting of the American Association for the Advancement of Science in Chicago on February 16, 1987.

Lear, John. 1978. *Recombinant DNA—The Untold Story.* New York: Crown Publishers. A lively account of the discovery and its aftermath.

Oren, Dan A. 1986. *Joining the Club: A History of Jews and Yale.* New Haven: Yale University Press.

Pais, Abraham. *Subtle Is the Lord . . . : The Science and the Life of Albert Einstein.* Oxford: Clarendon Press.

Watson, James D. 1980. *The Double Helix: A Norton Critical Edition.* Edited by Gunther S. Stent. Includes the text, commentary, numerous reviews, and some original papers. New York: W. W. Norton & Co.

General

Alberts, B., et al. 1983. *Molecular Biology of the Cell.* New York: Garland Publishing, Inc. A comprehensive textbook of cell biology.

Annual Reviews of Biochemistry. Palo Alto: Annual Reviews, Inc. Each volume contains about thirty authoritative reviews of special subjects. The 1988 volume, for example, has reviews on DNA repair and replication and the DNA polymerase III holoenzyme, in particular.

Darnell, J., et al. 1986. *Molecular Cell Biology.* San Francisco: W. H. Freeman & Co. The most current textbook of cell biology.

Kornberg, A. 1974. *DNA Synthesis.* San Francisco: W. H. Freeman & Co. My first monograph on DNA replication.

——— 1980. *DNA Replication.* San Francisco: W. H. Freeman & Co. Extended and revised sequel to *DNA Synthesis* (1974).

——— 1982. *Supplement to DNA Replication.* San Francisco: W. H. Freeman & Co. An updating of *DNA Replication.*

Olby, Robert C. 1974. *The Path to the Double Helix.* Seattle: University of Washington Press. A historical analysis of developing knowledge of macromolecules up to the discovery of the DNA structure.

Stenesh, J. 1975. *Dictionary of Biochemistry.* New York: Wiley-Interscience. A useful glossary.

Stryer, Lubert. 1988. *Biochemistry.* 3rd ed. New York: W. H. Freeman & Co. The most readable current textbook of biochemistry.

Watson, James D., et al. 1987. *Molecular Biology of the Gene.* 4th ed. 2 vols. Menlo Park: Benjamin/Cummings Publishing Co. A comprehensive and attractive textbook.

Glossary

A. Symbol for *adenine* or one of its derivatives.

Adenine. A member of the class of compounds known as *purines*; one of the key constituents of RNA and DNA; the key component of adenosine triphosphate (ATP) and some coenzymes (for example, NAD, coenzyme A).

Adenosine. *Adenine* with a ribose sugar attached.

Amino acids. The fundamental building blocks of proteins. Twenty different amino acids (for example, glycine, tryptophan, glutamic acid) commonly occur in proteins, in which they are linked to form chains several hundred units long; the sequence of the amino acids in such a chain is derived from DNA via RNA.

Bacteriophage. Any virus that infects bacterial cells. Such a virus consists of a nucleic acid (RNA or DNA) enclosed in a protein coat. In the smallest of these viruses, the nucleic acid (the chromosome) contains less than ten genes; in the largest, there may be more than 200. (For comparison, there are about 5,000 genes in a typical bacterium.) Like other viruses, a bacteriophage can survive, but cannot reproduce, outside a living cell. The nucleic acid, insinuated into a bacterial cell, directs the assembly of hundreds of new infectious particles, which are released in less than an hour. (Also sometimes referred to simply as "phage.")

C. Symbol for *cytosine* or one of its derivatives. Also the symbol for carbon, the most abundant element in living things.

Chromosome. A discrete, highly folded length of single- or double-stranded DNA. Nucleated cells (eukaryotes) have many pairs of chromosomes; nonnucleated cells (prokaryotes, or bacteria) have only one. The DNA in chromosomes is generally complexed with proteins (histones in eukaryotes).

Cytosine. A member of the class of compounds known as *pyrimidines*; one of the key constituents of RNA and DNA.

DNA (deoxyribonucleic acid). The substance of which genes and chromosomes are made. Information in DNA is transcribed into RNA for translation into proteins. DNA is distinguished from RNA in that it contains the sugar deoxyribose (rather than ribose) and the pyrimidine thymine (rather than uracil).

E. coli (Escherichia coli). A common bacterium of the intestinal tract. It is one of the favored experimental organisms ("guinea pigs") of genetics and biochemistry.

Electron. An elementary (subatomic) particle that has a negative electrical charge and orbits around the nucleus of an atom. An electron has roughly 1/1800 the mass of a *proton*.

Enzyme. A protein molecule that catalyzes biochemical reactions. Virtually all biochemical reactions in the body are catalyzed and directed by enzymes.

Fermentation. Metabolic breakdown of organic compounds that occurs in the absence of oxygen and yields a modest amount of energy. Examples are the conversion of sugar to alcohol and CO_2 in a yeast cell, or the conversion of sugar to lactic acid in a muscle cell deprived of oxygen.

G. Symbol for *guanine* or one of its derivatives.

Gene. A stretch of DNA at a specific location on the chromosome, generally encoding a specific protein. An average-sized gene of 900 nucleotides (or base pairs) encodes a protein with 300 amino acids.

Guanine. A member of the class of compounds known as *purines*; one of the key constituents of RNA and DNA.

Guanosine. *Guanine* with a ribose sugar attached.

H. Symbol for the element hydrogen. The heavy isotope, deuterium, makes "heavy" water; the radioactive isotope is tritium (^{3}H). Hydrogen is the simplest and most abundant element in the universe.

Lysis. Rupture and dissolution of cells.

Monomer. The repeating unit in a *polymer* (for example, amino acid in a protein, nucleotide in a nucleic acid, sugar in starch).

N. Symbol for the element nitrogen. It constitutes nearly ⅘ of the air by volume and is a constituent of all proteins and nucleic acids.

Nucleotide. The fundamental unit of RNA and DNA chains. Each nucleotide consists of a purine or pyrimidine linked to a sugar (ribose or deoxyribose), which in turn is linked to a phosphate group.

O. Symbol for the element oxygen. It is found in air as a pair of atoms (O_2), in water as one atom combined with two hydrogens (H_2O).

P. Symbol for the element phosphorus. In living things, phosphorus is nearly always combined with oxygen as a salt of *phosphate*.

φX174. A tiny *bacteriophage* with a chromosome of single-stranded DNA 5,386 nucleotides long encased in a roughly spherical (20-faceted) protein coat.

Phosphate. The molecule in which phosphorus is linked to four oxygen atoms ($PO_4^{=}$). When three hydrogens neutralize the negative charges, it is phosphoric acid; when a metal (such as sodium, potassium, or calcium) neutralizes the charges, it is a phosphate salt.

Plasmid. A small circular molecule of bacterial DNA (generally about 5,000 base pairs long) that replicates independently of the bacterial chromosome. Plasmids may carry genes for resistance to antibiotics, cleavage of DNA, and transfer of the plasmid from one cell to another.

Polymer. A high-molecular weight compound (for example, protein, starch, nucleic acid) consisting of long chains of repeating units called *monomers*, which may be identical or different.

Precipitate. The solid material obtained from a solution by altering the conditions (such as temperature) or by adding a specific substance (for example, salt).

Primer. A short chain of DNA or RNA, base-paired to the template DNA strand, which is elongated by DNA polymerase. The 3-end of the primer accepts the newly added nucleotide (building block) and is the starting point for DNA synthesis.

Proton. An elementary (subatomic) particle of the atomic nucleus with a positive electrical charge; identical to the nucleus of the hydrogen atom.

Purine. A class of compounds that includes the *adenine* and *guanine* building blocks of RNA and DNA.

Pyrimidine. A class of compounds that includes some of the building blocks of RNA and DNA (*uracil* in RNA, *thymine* in DNA, and *cytosine* in both).

Replication. Synthesis of DNA in which the two preformed (parental) strands of DNA serve as templates for new daughter strands (replicas).

Replication fork. The position on the DNA where replication is taking place. The parental strands diverge at this point to serve as templates for each of the newly synthesized (daughter) strands, thus creating a Y-shaped form.

RNA (ribonucleic acid). The polymer carrying the genetic message from DNA for the synthesis of proteins. RNA is distinguished from DNA by its content of the sugar ribose (rather than deoxyribose) and the pyrimidine uracil (rather than thymine).

Substrate. The substance (molecule) acted upon by an enzyme; its conversion to a particular product is catalyzed by a specific enzyme.

T. Symbol for *thymine* or one of its derivatives.

T2, T4. *Bacteriophages* with a large double-stranded DNA chromosome packed into the head of an elaborate structure containing a neck, body, and tail.

Template. In DNA synthesis, the DNA strand determining which building block (nucleotide) is to be added in turn to the primer strand.

Thymine. A member of the class of compounds known as *pyrimidines*; one of the key constituents of DNA.

U. Symbol for *uracil* or one of its derivatives.

Uracil. A member of the class of compounds known as *pyrimidines*; one of the key constituents of RNA.

Virus. The simplest life form, consisting of genetic information (DNA or RNA) packed into a protective protein coat, sometimes with an outer lipid envelope. In order to reproduce, a virus must enter a host cell, whose energy resources, building blocks, and, in some cases, replication and gene-expression machinery it commandeers. Bacterial viruses are called *bacteriophages.*

Vitamin. A small molecule, needed in tiny amounts for metabolic processes, that a particular organism cannot make for itself and so must obtain from the diet.

Index

A. See Glossary
Academic-industrial relations, 291–296
Aconitase, 51–56
Adenine, 96. See also Glossary; Purine nucleotides
Adenine ribose phosphate. See Adenosine monophosphate
Adenosine, 62. See also Glossary
Adenosine diphosphate (ADP): in RNA synthesis, 149–150
Adenosine monophosphate (AMP), 73; structure, 62, 70; synthesis, 142
Adenosine triphosphate (ATP): metabolic role of, 60–65; NAD synthesis and, 75–78; in nucleotide synthesis, 142, 143; production, 41–43, 45, 47, 49, 63, 64, 65; in protein synthesis, 82–83; role in DNA synthesis, 243; structure, 62; in synthesis of PRPP, 134, 135, 136, 138–139. See also Oxidative phosphorylation
Adler, Julius, 155, 173
ADP. See Adenosine diphosphate
Aerobic metabolism. See Respiration
Agar-agar, 116
Aging, 268
Alcohol fermentation, 32, 42, 59, 111–112; prize for explanation of, 30–31
Alkaptonuria, 83–84
Ames, Bruce, 82, 145
Amino acids, 37, 82–83. See also Glossary
Ammonium sulfate fractionation, 52, 76–77, 81–82, 234–235, 257
AMP. See Adenosine monophosphate
Amylase. See Starch
Anaerobic fermentation. See Fermentation
Andreopoulos, Spyros, 200, 202
Anemia, 17–20
Animal gene: introduction into E. coli, 282–283
Anthrax, 114–115, 187
Antimutator mutants, 210
Anti-Semitism, 310–314
Arai, Ken-ichi, 246
Arber, Werner, 279

ARP (adenine ribose phosphate). See Adenosine monophosphate
Ascorbic acid, 12, 145
Aspartic acid, 125–127
ATP. See Adenosine triphosphate
A-T polymers, 163–169
A-T tailing method, 276–277, 280
Avery, Oswald T., 151, 172, 272, 273; quoted, 302

Bacillus subtilis, 156, 186–187
Bacteria: cell cycle of, 254; cell extract preparation, 231–233; as experimental animals, 18; growth of, 115; size of, 190; soil, degradative abilities of, 100–101; spores of, 119–120; useful properties of, 104. See also particular species
Bacterial transforming factor, 156
Bacteriophage. See particular species; Glossary; Virus
Baldwin, Robert, 181, 182, 207, 277
Bambara, Robert, 247
Bangham, Alec, 218
Barker, H. A., 101–102, 125, 128, 172
Barley malt, 36–37
Barlow, Thomas, 12
Beadle, George W., 20, 84, 165
Beadle, Muriel, 165
Beckman, Arnold, 45
Beijerinck, Martinus W., 105
Berg, Millie, 138
Berg, Paul, 82, 138, 141, 142, 173, 174, 178, 179, 181, 182, 184, 230, 265, 277, 278
Beriberi, 9–11, 54
Bernard, Claude, 271
Berry, George Packer, 4, 189
Berthelot, Pierre, 35
Bertsch, L., 182, 184
Berzelius, Jöns Jakob, 31, 33, 112
Bessman, Maurice, 155, 173, 178, 210
Bigwood, E. J., 47
Biochemistry, 4–5; origins of, 269–274; unity of, 64–65, 242
Biotechnology, 289. See also Genetic engineering

Bloch, Konrad E., 131
Blood clotting, 20–24
Bloor, Walter R., 4, 173
Bodmer, Walter, 156
Botulism, 187
Boulgakov, N. A., 192
Boyer, Herbert, 281–282
Brain chemistry, 287
Brenner, Sydney, 262
Broker, Tom, 182, 276
Bronfenbrenner, Jacques, 313
Brutlag, Douglas, 184, 207, 208, 230
Buchanan, John, 97, 99–100, 122, 142
Buchner, Eduard, 34–35, 37, 60; quoted, 344
Bunim, Joseph, 50

C. *See Glossary*
Cairns, John, 217, 273
Cairns mutant, 217–220
Calcium: role in blood clotting, 24; role in transformation, 282
Cancer, 267–268
Carbamyl aspartic acid, 125–127
Carbohydrate synthesis, 83
Carbonic anhydrase, 298
Carter, Herbert E., 131
Catenanes, 278
Cell cycle: in animals and plants, 267–268; in bacteria, 254; cancer and aging and, 266–268
Cell-free replication system, 256–257
Cell membranes, 262–266
Cell wall, 262–263
Chain, Ernst, 314–315
Chamberlin, Michael, 230
Chargaff, Erwin, 272
Chemistry: medical science and, 285–289; understanding life as, 283–289
Cholera, 118–119
Chromatography, 136
Chromosome: defined, 92; of *E. coli*, 255; initiation of replication, 253–262; size, 89, 92, 93. *See also Glossary*
Citric acid cycle, 42, 54
Clarke, Hans T., 313
Clone: defined, 116
Clostridium botulinum tetani, 187
Clostridium perfringens (gas gangrene), 186
Coenzyme, 53; defined, 30; synthesis, 74–82, 142
Cohen, Seymour S., 271
Cohen, Stanley N., 281–283
Cohn, Ferdinand, 186
Cohn, Melvin, 178, 179, 181, 183
Cohn, Waldo E., 136

Colowick, Sidney, 72, 302; quoted, 66
Column chromatography, 136–137
Complementation, 233–234
Conditional mutants, 217–218, 234
Consumption. *See* Tuberculosis
Cooke, Alistair: quoted, 204–205
Cori, Carl F., 3, 30, 49, 65–69, 153, 179, 181, 313; laboratory of, 65–70
Cori, Gerty T., 3, 30, 65–69, 129, 153, 181; quoted, 273
Corner, George W., 4
Cozzarelli, Nicholas, 182, 196
Crick, Francis, 95, 121, 139, 160, 224, 272
CTP. *See* Cytidine triphosphate
Cuzin, François, 262
Cyanide, 43, 45, 47
Cytidine triphosphate: synthesis of, 142
Cytochromes, 42
Cytochrome c, 43, 45, 47
Cytochrome oxidase, 43, 47
Cytosine. *See also* Pyrimidines
Cytosine nucleotide: modification of, 153

Daft, Floyd, 71
Dam, Henrik, 20–21
Darwin, Charles, 306
Davis, Bernard D., 8, 131
Davis, Ron, 184, 281
DeKruif, Paul, 1
Delbrück, Max, 272–273
DeMars, Robert, 179
De novo pathway, 142–143
Deoxyribonucleic acid. *See* DNA; *Glossary*
Deuterium, 98, 99
Deutscher, Murray, 182, 211
d'Hérelle, M. F., 192
Dialysis, 61
Diastase, 37. *See also* Starch
Dicoumarol, 21, 23–24
Dieckmann, Marianne, 184
Discontinuous synthesis model, 225–226, 249–252
Disease: hereditary enzyme, defects and, 83–87, 144. *See also specific disease*
Djerassi, Carl, 174, 181
DNA (deoxyribonucleic acid): breaks in, 163–164; chain polarity, 161, 162; components of, 95–100, 147, 148; functions, 88–89, 94; GC pairs in, 163; isolation of, 146–147; molecular size, 38; nearest-neighbor analysis, 160; organization in chromosome, 175; as primer, 152–154; recombinants, 275–281; structure, 121, 158, 161, 162,

DNA (deoxyribonucleic acid) *(cont.)*
211, 213, 272; as template, 152, 153,
159; thymidine incorporation, 147–
148, 151–154; transfer between cells,
195, 266, 281–282. *See also Glossary;*
DNA repair; DNA synthesis
dnaA protein, 256, 259, 260, 261, 265–
266
dnaB protein, 243, 260, 261
dnaC protein, 243, 260
dnaG protein, 234, 237, 242. *See also*
Primase
dnaT protein, 243
DNA ligase. *See* Ligase
DNA polymerase: assay, 232; discovery
of, 147–154; functions of, 196, 207–
216, 229; and inability to start chains,
220; nuclease activity, 207–216;
proofreading activity, 208–216, 248;
purification of, 154–163; repair activity of, 213–216; role in replication of,
217–220; single-stranded templates
for, 194–196; of T4 phage, 210. *See
also* DNA polymerase I
DNA polymerase I, 219–220, 250–251,
276. *See also* DNA polymerase
DNA polymerase II, 219
DNA polymerase III, 219, 237, 238, 239,
275
DNA polymerase III holoenzyme, 239,
240, 247–252; subunits of, 249
DNA repair, 213–216
DNA Replication (Kornberg), 303–304
DNase (deoxyribonuclease), 151
DNA synthesis, 240–268; in broken-cell
extract, 146–154; chain growth during, 224–226; editing, 208–216; of infectious viral DNA, 197–206; initiation of, 220–239, 242–243, 253–262;
proofreading, 208–210; RNA primer
for, 229–231; temperature dependence, 167–168; tests of, 158–161; viral, 191–192, 197–206, 220–223, 228–
231; without template, 163–169
DNA Synthesis (Kornberg), 303
DNAX Research Institute, 304
Doudoroff, Michael, 81
Dubos, René: biography of Pasteur by,
110
Dyer, Rolla, 6, 7

E. coli. *See Escherichia coli; Glossary*
EcoR1 restriction nuclease, 278, 280,
281, 282
Edsall, John, 158
Eijkman, Christiaan, 10–11
Einstein, Albert, 312–313; quoted, 288–
289

Eisen, Herman, 180
Electron. *See Glossary*
Electron transport, 43–45
Elvehjem, Conrad, 17
Energy, 56–57, 60–65
Enrichment culture technique, 100–102,
123
Entropy, 56–57
Enzyme: activity, unit of, 55; amounts
required, 39; "dirty," 48–52, 66, 199;
function of, 36–41; hereditary defects
in humans, 83–87, 124, 144; history of
discovery of, 30–31; hunters, 29–58;
importance of, 298–299; intracellular
relationships of, 74; from one gene,
83–87; origin of term, 36; properties
of, 37, 39–41; purification, methodol-
ogy and advantages of, 48, 51–52,
299–301; reversible reactions and,
122–123; Section, NIH, 128;
specificity, 41; structure of, 37. *See
also Glossary*
Ernster, Lars, 274
Escherichia coli (E. coli): cell cycle, 254;
chromosome, 194, 255; culturing of,
156–157; DNA polymerases in, 219;
DNA synthesis in, 231; DNA transfer
into, 195, 281–282; extracts of, 147,
150; membrane of, 262–264;
minicells, 316; nutritional requirements of, 18; phage of, 89–93; replication origin in, 254, 255, 256, 257–
261; size of, 38, 194; transfer of
animal genes into, 282–283. *See also
Glossary*
Exonuclease III, 275, 276

FAD. *See* Flavin adenine dinucleotide
Falaschi, Arturo, 176
fd phage, 195
Fenn, Wallace O., 4
Fermentation. *See Glossary;* Alcohol fer-
mentation; Lactic acid fermentation;
Yeast
Fernandes, José, 139, 141
Feynman, Richard, 89
Flavin adenine dinucleotide: cleavage
and synthesis, 72, 73, 80–81
Fleming, Alexander, 233, 314
Florey, Howard, 314–315
Flory, Paul, 181
Fogarty, John E., 131
Folic acid, 8, 14, 19–20, 21–22
Franklin, Rosalind, 306
Friedkin, Morris, 147–148
Fruton, Joseph, 312; quoted, 159
Fuller, Robert, 256
Funk, Casimir, 15, 17

G. See *Glossary*
Gama, Vasco da, 11
Garrod, Archibald, 5, 83–84
GATC sequence, 259
Gefter, Malcolm, 219
Gellert, Martin, 196
Gene: hunters, 269–283; origin of term, 84. See also *Glossary*
Genentech, 283
Genetic engineering, 274, 275, 281–283, 289–296; controls on, 290–291; significance of, 284
Germination, 187
Germ theory of disease, 110–119
Gilbert syndrome, 5
Glucose: metabolism, 32, 41–43, 64; structure, 32
Glutamic acid, 20
Glycine: role in adenine synthesis, 99
Glycogen phosphorylase, 67, 153
Godson, Nigel, 237
Goldberger, Joseph, 8, 15–17
Goodgal, Sol, 156
Goulian, Mehran (Mickey), 182, 196–197, 204–205
Gout, 86, 144, 145
G4 phage, 237–239
Green, David E., 49, 123
Greenberg, G. Robert, 100, 122, 142
Greenwald, Isidor, 50
Griffith, Fred, 272
Grunberg-Manago, Marianne, 149
Guanine. See *Glossary*; Purine nucleotides
Guanosine. See *Glossary*

H. See *Glossary*
Haas, Erwin, 49
Harden, Arthur, 60, 61
Hartwell, Jonathan, 173
Hawkins, Richard, 11
Hayaishi, Osamu, 101, 123, 128, 179
Heidelberger, Michael, 172
Helicase, 245, 249–250
Hemophilus influenzae transforming factor, 156
Hemorrhagic disease in cattle, 23
Heppel, Leon A., 7, 12, 71–72, 79, 128, 129
Hereditary disease, 5–6, 83–87, 144, 145
Hershey, Alfred, 273
Hevesy, Georg von, 97–99, 173
HGPRTase. See Hypoxanthine guanine phosphoribosyl transferase
Hill, J. Lister, 131
Hirota, Y., 256
His, Wilhelm, 271
Hoefer, Peter, 183

Hogness, David, 178, 179, 181, 182, 184
Holmes, Oliver Wendell, 113
Hook, Sidney, 309
Hopkins, Frederick Gowland, 12, 13–15, 19, 313; quoted, 270
Horecker, Bernard L., 43, 45, 47, 49, 71–72, 79, 128, 129
Horowitz, Norman, 217
Hotchkiss, Rollin, 156
Hurwitz, Jerard, 178, 179, 196, 235, 236
Hydrogen: isotopes of, 98–99
Hypoxanthine guanine phosphoribosyl transferase, 85–87, 142, 144

Insulin: amino acids in, 39
Invertase, 35, 37
Ion exchange chromatography, 136
Isotope: defined, 98
Isotope tracer technique, 97–99

Jackson, David, 278, 280
Jacob, François, 179, 262
Jaundice, 5–6
Johnson, Lyndon B., 202–203
Joining the Club: A History of Jews and Yale (Oren), 312
Josse, John, 159–160

Kaguni, Jon, 256, 257
Kaiser, A. Dale, 178, 179, 181, 182, 184, 276, 277, 278, 280
Kaiser-Hogness technique, 276, 282
Kalckar, Herman, 30, 50, 65–66, 69, 147, 148; quoted, 66
Kaplan, Henry, 172, 180
Kaplan, Leah, 172
Keilin, David, 49, 172
Kendall, E. C., 12
Kennedy, Edward, 290
Kennedy, Eugene, 235, 265
Kennedy, Joseph, 298
Khorana, Gobind, 138–139, 166
Kinase, 143
King, C. Glen, 12
Kleiner, Alfred, 312
Kluyver, Albert J., 101, 105
Koch, Robert, 10, 15, 114–116, 117–119, 187, 300
Koch's postulates, 115
Korn, Edward, 82
Kornberg, Arthur: on administrative work, 304–305, 308, 317–318; on basic research, 314–316; birthday party at 70, 154; childhood of, 2–4; in Navy, 2, 6, 7; education of, 2, 4, 24, 56–57, 105–106, 117, 189, 310–311; on enzyme research, 74; on exchange of research information, 235–236; family

Kornberg, Arthur *(cont.)*
  and residences of, 3–4, 50–51, 66,
  105, 170–177, 186; on life as chemistry, 283–289; photograph of, 128, 170,
  178, 182; on prejudice, 309, 310–314;
  prizes won by, 82, 171; publications
  and speeches of, 5, 22, 47, 81, 82, 155,
  158, 159, 200–201, 230, 284, 288, 289,
  298, 302–304, 305, 307, 316; on teaching, 301–302, 305–306, 317; on time,
  79, 301–302; on writing, 302–304
Kornberg, Ken, 102, 170, 173, 176–177
Kornberg, Roger, 66, 170, 171, 173–175,
  230
Kornberg, Sylvy, 3, 47, 141, 155, 171–
  173, 177, 230; quoted, 172
Kornberg, Tom, 79, 170, 173, 175–176,
  218, 219
Kuffler, Stephen, 172
Kühne, Wilhelm, 36
Kunitz, Moses, 151

Laboratory design, 176–177
Lactic acid fermentation, 42, 59
Lactic dehydrogenase: purification, 67,
  69
Lactobacillus casei, 19
Lagging strand, 225, 226, 250, 251
Lambda (λ) exonuclease, 275, 276
Lambda (λ) phage, 196, 276
Lavoisier, Antoine, 31–32
Lawrence, Peter, 176
"Lazarus effect," 164
Leading strand, 225, 226, 250, 251
Lebedev, Alexander, 75
Lebedevsafts, 75. *See* Yeast extract
Lecithin, 264
Lederberg, Joshua, 156, 181, 282. *See
  also* Foreword
Lehman, I. Robert, 151, 153, 154, 155,
  164, 173, 178, 179, 181, 182, 184, 196,
  210, 276, 277
Lehninger, Albert L., 71, 131
Lesch-Nyhan syndrome, 85–87, 145
Lieberman, Irving, 124–125, 128, 134,
  141, 142, 179
Liebig, Justus von, 31, 35, 112
Life: "creation" of, 201–206
Ligase, 196, 197, 215, 251, 275, 276, 281
Lind, James, 11
Lindberg, Olov, 69–70
Lipmann, Fritz, 30, 69
Littauer, Uri, 147, 149
Liver extract, 18, 78, 100, 134–135
Lobban, Peter, 275–278, 280–281
Lobbying, 295–296
Low, Bob, 246
Lowry, Oliver, 72, 179, 302

Luria, Salvador, 179
Lynen, Feodor, 173
Lysis. *See Glossary*
Lysozyme, 233

Madden, Sidney, 311
Malic enzyme: purification, 57
Maltase, 80
Mandel, Mort, 282
McCann, William S., 6, 311
McCarty, Maclyn, 151
McCollum, Elmer V., 15
McConnell, Harden, 174, 181, 218
McHenry, Charles, 247
Mehler, Alan, 57, 69
Membranes: role in cell division, 262,
  266; role in DNA synthesis, 218, 228–
  231, 265–266; structure of, 174–175,
  262–266
Mendel, Gregor, 84
Mertz, Janet, 281
Meyerhof, Otto, 30, 49
Microbe hunters, 106–120
Microbiology (Davis et al.), 180
Microscope, early, 107
Miescher, Friedrich, 271, 291
Minicells, 316
Mitochondria, 71
Mitra, Sankar, 195
Molecular biology: origins of, 269–274
Molecular size, 37–39, 190
Monoclonal antibody technique, 284
Monod, Jacques, 179
Monomer. *See Glossary*
Morales, Hilbert, 178, 183
Morrow, John, 280–282
Morton, Robert, 78
M13 phage, 195, 228–231
Mutator mutants, 210
Mycobacterium tuberculosis, 118

N. *See Glossary*
NAD. *See* Nicotinamide adenine dinu-
  cleotide
NADP. *See* Nicotinamide adenine di-
  nucleotide phosphate
NAD synthetase: isolation and
  purification, 75–82
Nathans, Daniel, 279
National Institute(s) of Health (NIH):
  history of, 16, 22, 71–72, 129–133,
  294; A. Kornberg at, 2, 6, 7–8, 17, 22,
  49, 79, 124, 128, 297–298
Nearest-neighbor analysis, 160–161
Needham, John, 108
Neuberger Report on Food and Nutrition
  Research, 27
Neurospora crassa, 30, 84, 124

Niacin: assay, 54; in coenzyme, 53; deficiency, effects of, 54
Nicotinamide adenine dinucleotide (NAD): breakdown, 77–78; cleavage, 70, 72–73; reduced, in orotic acid breakdown, 125; structure, 70; synthesis, 75–78, 97
Nicotinamide adenine dinucleotide phosphate (NADP), 53–55; function of, 53–54; structure, 54, 73–74, 136; synthesis, 81
Nicotinamide ribose phosphate (NRP), 73; in NAD synthesis, 75–78, 80; structure, 70
Nicotinic acid. *See* Niacin
Nightingale, Florence: quoted, 113
NIH. *See* National Institute(s) of Health
Nobel Prize, 171–172, 274
NRP. *See* Nicotinamide ribose phosphate
Nuclease activity, 207–216
Nucleic acids: differences in, 147, 148, 229, 272; discovery of, 271–272; synthesis of, 83. *See also* DNA; RNA
Nuclein, 271
Nucleoside diphosphates: in RNA synthesis, 149–151
Nucleosome, 175
Nucleotide pyrophosphatase, 73
Nucleotides, 96–97, 272; chromatographic separation, 136; defined, 92; in DNA synthesis, 152; synthesis of, 142. *See also* Glossary; Purine nucleotides; Pyrimidine nucleotides; *specific nucleotide*
Nutrition: decline of, 8, 24–28; as a science, 13–15

O. *See* Glossary
Ochoa, Severo, 30, 46, 49–50, 51, 57–58, 69, 129, 149–150, 172
Okazaki, Reiji, 223, 225–226; fragment, 226, 250; maneuver, 223
Okazaki, Tuneko, 223
Olby, Robert C., 273
Oligomer: defined, 167
Oncogene, 74, 268
One gene, one enzyme concept, 83–87
Oren, Dan A., 312
oriC, 254, 255, 256, 257–261
oriC plasmids, 254, 256, 259, 265
Orotic acid, 123–127; degradation of, 101–102; in DNA synthesis, 127, 134–136; in nucleotide synthesis, 143; structure, 124
Oxidation: defined, 42–43
Oxidative phosphorylation, 49, 65, 69
Oxygen: metabolic role of, 41–43, 44, 45, 60

P. *See* Glossary
PABA. *See* para-Aminobenzoic acid
Pais, Abraham, 312
Palade, George, 218
para-Aminobenzoic acid, 21, 22
Parran, Thomas, 129
Pasteur, Louis, 15, 31, 33, 35, 105, 108, 109–112, 300; quoted, 33, 35, 110
Pasteurization, 112
Pauling, Linus, 272, 306; *Nature of the Chemical Bond*, 56
Pellagra, 8, 15–17, 54
Penicillin, 263, 314
Pepsin, 37
Periplasm, 262, 263
Perutz, Max, 205–206
Peters, Rudolph, 49
Petri dish, 116
Phage. *See particular species;* Virus
Phage and the Origins of Molecular Biology (Cairns, Stent, Watson, eds.), 273
φX174 phage, 192–200, 228, 234, 235, 237, 242–245; in vitro synthesis of DNA of, 197–206; life cycle, 244; mutator and antimutator mutants of, 210; size, 190. *See also* Glossary
Phosphate: in DNA, 95, 96; role of, 61–64. *See also* Glossary
Phosphatidic acid, 264
Phospholipid, 264–266; bilayers of, 174–175; synthesis, 83
5-Phosphoribosyl-1-pyrophosphate (PRPP), 138–141; in nucleotide synthesis, 142–143
Pigeon liver, 51–58; extract, 135–136
Pig heart muscle, 51–58
Plaque: defined, 193
Plasmid, 241–242, 266, 281. *See also* Glossary
Pneumococcus pneumoniae transforming factor, 156
Pol I. *See* DNA polymerase I
Pol II. *See* DNA polymerase II
Pol III. *See* DNA polymerase III
Polyethylene glycol, 257
Polymer. *See* Glossary
Polynucleotide phosphorylase, 149–151
Polyphosphate, 141
Potato enzyme, 72–73
P22 phage, 280
Precipitate, 53, 146. *See also* Glossary
Prelog, Vladimir, 139
Pricer, William E., 79, 128
Primase, 243. *See also* dnaG protein
Primer, 152–154. *See also* Glossary
Primosome, 240, 242–246, 250, 251–253. *See also* DNA synthesis, initiation of

Programmed cell death, 268
Protein chain: defined, 37; insulin, 39
Protein content: assay of, 55–56
Proteins n, n', n", 243
Protein synthesis: role of DNA in, 94
Proton. *See Glossary*
Proto-oncogene, 268
Protoplasm, 15
PRPP. *See* 5-Phosphoribosyl-1-pyrophosphate
PRPP synthetase, 138
Pterin, 19
Puerperal fever, 113
Pure culture technique, 115–116
Purine. *See Glossary*
Purine nucleotides, 122; synthesis of, 142–144
Pyrimidine nucleotides, 122; precursors of, 127, 134; synthesis of, 142–144
Pyrimidines: biosynthesis and degradation, 123–124. *See also Glossary*
Pyrophosphate, 142; inorganic, 69, 70, 77–78; protein synthesis and, 82–83
Pyruvic acid: oxidation, 42–43, 49

Rabbit kidney enzyme, 69–70, 72
Racker, Efraim, 69; quoted, 48
Radium: decay products, 97–98
Recombinant DNA: production of, 275–281; technique, 275–283
Redi, Francesco, 108
Reich, Edward, 211
Reiterative replication, 167–169
Repair. *See* DNA repair
Replication: defined, 94; reiterative, 167; rolling circle, 245. *See also Glossary*; DNA synthesis
Replication fork, 224–226, 249–253. *See also Glossary*
Replisome, 240–241, 251, 252, 253
Resistance: to antibiotics, 281–283
Respiration: defined, 41; in vitro, 41–48
Restriction nuclease, 278–280
Retrovirus. *See* RNA virus
Reverse transcription, 95
Riboflavin phosphate, 80–81
Ribonucleic acid. *See* RNA
Ribose, 95, 96, 148
Ribose phosphate, 135, 136, 138, 139
Richardson, Charles, 174, 196, 219
Rickets, 13
Rifampicin, 230–231
RNA (ribonucleic acid): degradation and synthesis, 149–150; messenger, 94. *See also Glossary*
RNA polymerase, 151; functions of, 229; role in DNA synthesis, 229–231

RNA primer, 229–231, 250, 251; synthesis of, 236–237
RNase H, 261
RNA virus, 94–95
Roentgen, Wilhelm, 315
Rolling circle replication, 245
Rose, Leonard, 176

Saccharomyces cerevisiae, 103. *See also* Yeast
Salvage pathway, 143–145
Sanger, Fred, 305
Schachman, Howard, 163
Schekman, Randy, 235
Schmeck, Harold, 202
Schmidt, Gerhard, 264
Schneider, Morton, 57
Schrecker, Tony, 79
Schwann, Theodor, 33
Scurvy, 11–12
Sebrell, W. Henry, 8, 30, 49, 66, 71
Secrecy: in science, 292–293
Semmelweis, Ignaz Philipp, 113
Shannon, James A., 131
Shear, Murray, 173
Sherberg, Esther, 182, 183
Simian virus (SV) 40, 277–278, 280
Siminovitch, Lou, 280
Simms, Ernie, 138, 142, 155, 178, 181, 183
Single-strand binding protein, 237, 238, 243, 250
Sinsheimer, Robert, 193–194, 196
Smith, Emil, 312
Smith, Hamilton, 279
Snow, C. P., 288
Soddy, Frederick, 98
Sonic extraction, 157
Spallanzani, Lazzaro, 108–109
Spectrophotometer, 45, 46
Spontaneous generation theory, 108–109
Spore, bacterial, 119–120, 186–189
Sporulation, 187
Stadtman, Earl, 101
Stadtman, Terry, 101
Stanford Biochemistry Department, 177, 180–185
Stanley, Wendell, 163
Starch, breakdown, 36–37
Stark, George, 182, 183
Stent, Gunther, 273, 275
Sterilization, 186–187
Sterling, J. E. Wallace, 181
Stone, Edward Durell, 181
Stryer, Lubert, 182, 183
Substrate: properties, 39, 41; reaction with enzyme, 37, 39–41. *See also Glossary*
Succinic acid, 43

Succinic acid oxidase, 43, 47
Sucrase, 37, 39. *See also* Invertase
Sucrose: breakdown of, 39, 40; fermentation, 31–32; structure, 32
Sueoka, Noboru: quoted, 226–227
Sugar metabolism, 29–31
Sulfa drugs, 17–20, 21–22
Sulfonamides, 315
Svirbely, Joseph, 12
Symons, Bob, 278, 280
Szent-Györgyi, Albert, 12
Szent-Györgyi, Nellie, 12

T. *See Glossary*
Tabor, Celia, 171
Tabor, Herb, 79, 129, 171
Takaki, K., 9–10
Tamelen, Eugene van, 181
Tartaric acid: optical forms of, 111
Tatum, Edward L., 30, 84
Taube, Henry, 181
Taylor, John, 67
Temperature-sensitive mutants, 218, 234
Template, 152, 153, 159. *See also Glossary*
Terman, Frederick E., 181
Terminal transferase, 275, 276
Tetanus, 187
Thannhauser, S. J., 264
Theorell, Hugo, 47–48; quoted, 47
Thiamine, 11, 54; assay for, 18
Thomas, Lewis: quoted, 130
Thymidine: incorporation into DNA, 147–148, 151–154; synthesis, 147, 149
Thymidine dimers, 214–216
Thymine, 148. *See also Glossary*; Pyrimidines
Todd, Alexander, 139
Tomkins, Gordon, 82
Topoisomerase, 250, 261
T2 phage, 89–93, 273; DNA in, 153; mutator and antimutator mutants of, 210; size, 38, 190. *See also Glossary*
Transcription: defined, 94
Transformation, 282
Transforming factor (bacterial), 156
Translation: defined, 94
Traube, Moritz, 35
Tritium, 98, 99
Tropical sprue, 18
Tswett, Michael, 136
Tuberculosis, 117–118
Twort, F. W., 192
Tyndall, John, 186

U. *See Glossary*
Ultraviolet-induced damage, 213–216

UMP. *See* Uridine monophosphate
Uracil, 148; structure, 124. *See also* Pyrimidine
Uracil ribose phosphate: synthesis, 134–135, 140. *See also* Uridine monophosphate
Uric acid, 99–100, 145. *See also* Gout
Uricase, 145
Uridine monophosphate, 127, 140, 141

Van Deenen, Laurens, 218
Van Leeuwenhoek, Anton, 33, 105, 106–108
Van Niel, Cornelis B., 101, 105–106; quoted, 106
Vibrio cholerae, 119
Vinograd, Jerome, 199
Virus, 119–120, 189–192; chromosome size, 89; DNA synthesis for, 191–200, 220–223, 228–231; in vitro synthesis of infectious DNA of, 197–206; replication, 193; size, 38, 190. *See also Glossary*
Viscometer, 163
Vitalism, 33–34; in modern research, 300
Vitamin, 15; assays for, 18–19; metabolic functions, 29–30; prophylactic use of, 26. *See also Glossary*
Vitamin A, 15
Vitamin B. *See* Niacin
Vitamin B1. *See* Thiamine
Vitamin C. *See* Ascorbic acid
Vitamin D, 13, 15
Vitamin deficiency disorders, 8–24
Vitamin hunters, 1–28
Vitamin K, 20–24

Wallace, A. R., 306
Wallis, W. Allen, 311–312
Walpole, Horace, 34
Wang, James C., 278
Warburg, Otto, 30, 45, 48, 53, 54; quoted, 66
Warfarin, 21, 24
Washington University School of Medicine, Microbiology Department, 177–180; A. Kornberg at, 3, 30, 105, 119–120, 129, 146, 298, 301, 311, 313
Watson, James, 121, 160, 224, 272, 273; and *The Double Helix*, 302, 308; quoted, 299
Whipple, George H., 4, 6, 118, 310, 311
Wickner, Bill, 235
Wickner, Sue, 235
Wilkins, Maurice, 306
Wills factor, 18
Wine: diseases of, 111–112

Winogradsky, Sergei N., 105
Wöhler, Friedrich, 31, 33
Wolf, Arnold V., 6
Wood, Harland, 172, 179
Woolf, Esther, 66
Woolf, Ralph, 66
Woolley, D. Wayne, 17

X-rays, 315

Yanofsky, Charles, 172, 174, 181, 300
Yasuda, Sei-ichi, 256

Yeast, 103–104; fermentation in, 31–34, 41–42, 59, 60, 111–112; size, 190
Yeast extract, 14–15, 18, 81, 134–135; boiled, 60–61; enzymes in, 80; NAD synthesis with, 75–78; preparation of, 34–35, 75

Zaffaroni, Alex, 283
Zimmerman, Steven, 155
Zinder, Norton, 195
Zymase, 35, 37
Zymobacterium oroticum, 102